ESCAPE

A South African Breaks Out and Finds Peace

Jaime Salazar & Keryn Barnes

Johannesburg, South Africa

info@alexdebruyn.com

© 2017 Alex de Bruyn

FOREWORD

Our existence is a mosaic of stories woven together by design. Most tales go untold. Before joining the French Foreign Legion, I never understood why former members rarely spoke of their experiences. In 2013, still picking up the pieces of my time within this mysterious corps, my sister Keryn Barnes convinced me that this peculiar chapter of my life was worth retelling. Over the course of several weeks—replete with wine, tears, and joy—I poured my soul into a Dictaphone, a cathartic exercise in atonement and healing. She amassed my thoughts and built the basis of what would become this narrative. Soon after that, I met author Jaime Salazar, a fellow legionnaire who penned the acclaimed memoir *Legion of the Lost*. I thank him for not only overhauling my rough draft but for believing in my story and masterfully transforming my thoughts into compelling prose. His technical skill, creative intelligence, and craftsmanship were indispensable in this very personal endeavor.

The names of characters in this account have been changed. Creative license was used sparingly, and some details have been altered to protect the identity of those involved. Some occurrences were second-hand accounts. This retelling does not express a strict chronology, and specific events were merged for the sake of pace, succinctness, and clarity.

The depiction of my service in the Foreign Legion—which I hold in high esteem—describes individual experiences and interpretation of events at a particular time, and is not necessarily a portrayal of the Foreign Legion as a whole. This is my dispassionate but unvarnished story.

Alex de Bruyn, Johannesburg, 2018

Dedicated to every boy or girl who has a burning, curious desire for adventure fueled by the knowledge that there is more to the miracle of life than what is randomly placed before them.

Table of Contents

- FOREWORD 2
- HIPPO 5
- THE ELECT 28
- IRON SHARPENS IRON 55
- EDEN 89
- JOB 115
- ASH WEDNESDAY 142
- BEN-HUR 168
- PHILIPPI 185
- TOWER OF BABEL 213
- SODOM 229
- EXODUS 245
- ODYSSEUS 254
- LAND OF NOD 268
- THE PRODIGAL SON 277

HIPPO

Legionnaire: you are a volunteer serving France faithfully and with honor.

Code d'honneur

Caught in the notorious doldrums, we floated helplessly in the Atlantic—the Blue Amazon—near the equator, and it had been a week since we had had a mere puff of wind fill the raggedy sails of our damaged and leaking boat. *How did I get myself into this?* Three men were about to die of thirst, and it was my fault. Was the only unforgivable sin stupidity, or hubris? Pride comes before the fall, they say. I felt the piercing sun drying the water right out of my skin as if life was being slowly peeled from my helpless body. This was one of the most painful ways to go, especially with so much water around us. If I drank seawater, I'd end up pissing out more than I consumed to rid my corpse of the excess salt. *What if I sipped small amounts?* I resisted the temptation and lay stoically on the hard deck, on scorching planks that broiled me from below.

Above us were the vapor trails of airliners en route from Rio de Janeiro to Miami. *Save us!* My mind screamed, yet I could barely twitch a muscle. There was a sudden break in the sun's relentless radiance, which prompted me to again peel my encrusted eyelids from each other. A bird was circling us in wide berths, its intimidating wingspan casting a shadow. It was waiting for the sun to finish us off. *This is it. We won't end up as fish food but bird shit.*

"It's a Brown Booby," said Connor, our Australian *wunderkind* at the helm. "They're harmless. Their presence means that land might be closeby. Vultures don't make it out this far."

"How the hell do you know that?"

He didn't answer. I chalked it up to his mysterious and angelic wisdom. The bird was disinterested in our agony but simply hovered over us like a watchman. Connor was plopped at the cockpit with his legs up, one arm on the wheel and the other holding onto a dog-eared paperback, *The Kite Runner*, which he'd been reading to Vasile, a young Romanian, and me the entire morning. *"It may be unfair, but what happens in a few days, sometimes even a single day, can change the course of a whole lifetime..."* Such asides kept our minds away from death, which was crawling closer to us by the hour.

"Connor, I can't take the thirst much longer. Please pass me a little water," I asked as sticky white saliva formed strings between my cracked lips. "We only got a liter of fresh water left, among us three," he cautioned sternly. I became paranoid that Vasile would drink all of it while Connor and I slept, and blame me for the crime. Desire could turn any man into an animal. But then Connor chucked a bottle at me. "I was joking. Enjoy a good swig of mine. I'm not thirsty." I knew he was lying. He always sacrificed himself for others, but I wasn't about to let him die for my depraved soul.

"When do we drink our piss?" I asked boyishly. "Should we start saving it just in case?"

"Stop thinking about that," Connor muttered. I then picked up a container and relieved my bladder of viscous urine, which had the color, smell, and consistency of motor oil. Vasile was lying below deck, the only shaded area on the boat, to escape the blazing sun. We could roast out on the deck, or sweat out our last drops of water and suffocate in the hold.

Our forty-foot sloop had seen better days, perhaps when French Guiana gained independence. We'd been dogging the hell out of the vessel for weeks, but we loved her, despite her bruises and stitches. The hull was sleek and slender but had long ago turned brown from neglect. Below deck, frayed wires and broken instruments were scattered about. We had to take it or leave it if we wanted to get the hell out of Paramaribo after I'd been freed from jail. "You expect us to pay top-dollar for a leaking, gutted, Styrofoam cup with a rotting sail?" Connor protested to the local fisherman, but we had no other options. "Well then, I guess we'll take it!"

In a mad rush to come and save us from Paramaribo, Connor ran squarely into a shoal, taking on a lot of water. He barely avoided scuttling the vessel, but the flooding left it without any useable instrumentation. On our first attempt to leave Paramaribo harbor, a rope tangled around the spinning propeller and the packing gland snapped. All that now held it together and kept the boat from being swallowed by the Atlantic were two vice grips and four cable ties. For good measure, Vasile added his used chewing gum as a backup patch.

Vasile opened his eyes as I stepped down into the cockpit. "Which way we heading?" he asked.

"Still northwest," I responded. The Legion meticulously taught us to navigate according to the sun, stars, or landmarks, but those skills were useless if we couldn't control our direction. With the powertrain in ruins, we were entirely at the mercy of the Guiana Current, which was pushing us into the Caribbean, the wrong trajectory.

I avoided sitting at all costs to avoid the horrible spins and vertigo, but my emaciated legs eventually gave out. As I sat down, I immediately swung my body around, gripped onto the stern rail, and started retching. Every muscle in me clenched, and my stomach contracted like a bagpipe. A week of sickness left very little for my body to eject.

7

We'd gone through our provisions but didn't need much food anyway, for it only brought on thirst. We'd die from a variety of ailments before we starved. I'd gone days without a single calorie during the Legion's renowned Advanced Jungle Warfare Course. During the last days, we were allowed the delectable treat of fat, white tree grubs.

Finally, some yellow bile and blood made its way up, and I projectile-vomited it into the ocean, reminiscent of *The Exorcist*. My throat burned from the acid. My eyes were watering, and sweat ran from my forehead to my nose. I sat and simply stared into the water. "*Make it go away!*" I pleaded to whatever higher power was listening. I then felt an indescribable peace when the heaving stopped. But as I gazed deep into the water, I noticed three Great Whites appear from the depths. "You bastards smelled my blood and bile, huh?" The largest one darted directly towards me, staring deep into my bloodshot eyes. His dorsal fin broke the water and grazed against the hull as he joined the other two. I looked again and saw that the shark had a sick smile on his face.

"Get it over with. Jump in! Nobody will miss you," he mouthed at me.

"Hey wait," I said, snapping out of a trance. "You're not really a shark. I won't fall prey to your snares! Wicked one, get behind—"

"Hey, are you are fucking hallucinating again?" I heard Connor's voice crack above my head as he peered over the stern rail to see what nonsense I was blubbering about. I stared up at him, and when I looked back, there was nothing but crisp, clear water. I was slowly losing my mind.

"Hey mate, I don't know how much longer it'll be," he continued, "but you need to get your head straight. Vasile is losing it too. I need you to be strong. If I lose you, then I'll lose him." Connor then returned to the cockpit to continue reading to us where he left off.

8

I had considered studying law before entering university. One required reading was the nineteenth century English case of *R v Dudley and Stephens*, which established a precedent that necessity is not a defense for murder. Dudley and Stephens were shipwrecked along with two other men. Facing certain death, they conspired to kill the young cabin boy for food. In a bizarre twist, the survivors were rescued very shortly after, making the killing pointless. After a highly publicized trial, the defendants were convicted of murder and sentenced to death but with a recommendation for clemency. My worries didn't play with civil law. I wasn't sure if God would forgive murder or cannibalism even in extenuating circumstances. I'd rather die with an untainted soul. But would letting myself starve when there was an edible corpse available not be tantamount to suicide?

Christ wandered in the wilderness and was tempted with food and drink. Elijah also suffered in the desert, but angels presented him with a meal that allowed him to survive for 40 days and 40 nights. Was Connor one of those cherubs?

When the heat broke, and the sun rested for the evening, Connor cheerfully stripped naked and jumped into the ocean for his daily shower. Where he got the energy was anybody's guess. "Number 4 of the Legion's *code d'honneur*: 'Appearances must always be impeccable,' you smelly bastards." But sometimes, gallows humor was exactly what we needed. "Hey, we won't eat you," Connor shouted at me. "You've been gorging on nothing but curry this past month. You'll taste too gamey!"

Vasile laughed, rubbed his hands as if sitting before a buffet, and smirked, "I hunt wild boar in Romania. I like the gamey taste."

"How about you wankers share my big balls after I'm gone? Only one per person. Actually, Connor, you're as heavy as both of us. You're a better 2-for-1 value. We could go for months on your lean carcass."

After the guffaws, I got up from the stern and sat in the cockpit. The first shift would be mine while the lads got some shuteye. As the sun set, I sat back and let my eyes wander about the heavens. Millions of stars were scattered across the black canvas of the universe. Suddenly I heard distinct laughter in the distance, something like a child running through the darkness playing hide-and-seek. I jumped up to see who was there and scanned the horizon for something to convince me that I wasn't hallucinating again. The sound became a swooshing hum right beside the boat. I looked down and saw long florescent streaks through the water. They started on the sides of the hull, scooped under the boat, and emerged on the other side. It was beautiful as if the Northern Lights had dove into the Atlantic and presented themselves as a ray of hope. While I watched the ocean dance before me, I noticed that it was from a pod of dolphins breaking the water, displaying their brief but comforting bioluminescence.

"Rest now," Vasile announced as he emerged from below deck to take his turn. A Legion discipline that had stuck with us was making sure that someone always remained on watch. I then lay next to the mast on deck. I couldn't go below, as that was the quickest way to aggravate my seasickness. But just as I fell into blissful sleep, I suddenly felt my stomach heaving and blood rushing to my head. There was no time to gain my bearings. I simply rolled to the side and let my guts vomit on deck. "You better wash that off!" Connor shouted from his slumber. My eyelids dropped, and my mind finally rested.

When I woke, the sun was already high in the sky, burning me to a crisp. My anguish continued. I wiped the remainder of the bile from the corner of my mouth and felt my cracked lips grate against my hand. "Still no wind?" I pleaded as I tried to remember how many days had passed since this blur of pain had begun.

"Nothing," Vasile replied from the bow. I turned to see him standing with his feet locked into the bow rail; his hands stretched out to the heavens in a Jesus pose. "If there is a god, show yourself! Bring back the storms. I'd rather take your fury than your calmness. We're not scared!" he shouted at the sea. "Screw you. Bring it on!" He sobered up moments later. "I spent my childhood herding cattle in Transylvania, where winters get to -20 °C. I never thought I'd die under this cruel sun in the middle of the Atlantic."

"Connor deserves to live," I put in desperately. "He had a swimming career. He'll produce some tall, beautiful Australian children. Humanity needs him. Me? I'm bad with women, a failed business heir, and shitty rugby player. And now I'm a scrawny, sickly excuse of a man. Life had been good, but I had the stupid idea that I could make my family proud by joining this godforsaken Legion. They'll think I died heroically in the jungle, rather than as a cowardly deserter on some cursed boat. I wish I could take it all back."

"Both of you get your heads out of your asses," Connor snapped. "We're going to make it."

My lot wasn't meant to end like this. One's life flashes before one just before death. I stared at the sky as clouds gathered, and retraced the fateful events that got me to where I was at that moment. I reminisced about the good old days in my beloved Johannesburg. And just then, I felt a rain droplet hit my forehead.

I remember the first rugby match I watched with my stepfather, Pa. I sat calmly on his lap during the 1995 Rugby World Cup in my hometown. The competing pugilists were our very own Springboks and New Zealand's infamous All Blacks. My biological father never had the opportunity to take me to an international rugby match, for he died during apartheid, while South Africa was banned. My homeland was finally reintroduced into international sporting events after the fall of white rule in 1991. This World Cup was South Africa's chance to prove that it could change for the better and still be the best.

Being the underdogs, the Springboks started the match mainly on the defensive. Due to the strength of the flankers and the heart of scrumhalf, the All Blacks couldn't break their line. All subsequent points were scored by kicks, with the eventual armistice at an even 9 - 9 at the final whistle. "Overtime," Pa whispered in my ear as the throngs roared. I could feel the collective passion as he initiated me into manhood. "It goes to show," Pa said cheerfully, "it's never too late to redeem oneself."

Early into overtime, New Zealand took the lead, by slotting a long-range penalty, leaving South Africa expecting the worst. Time passed slowly, but at the last moment, the Springboks evened the score to 12 -12. With minutes to go, the Springboks got a scrum thirty meters from the All Black's tryline. Breaking it was impossible, so they hung back and waited for the perfect moment to strike. The ball was fed to the scrum and passed deep to the fly-half who was now comfortably in the pocket, lining up the goal. A mere second later, he dropped the ball to his foot. Defenders desperately dove to try to block the kick but failed.

As the ball floated in slow motion through the uprights, Pa lifted me high in the air. The Earth shook as 43 million of my countrymen jumped in unison as the underrated Springboks dethroned the mighty All Blacks.

I was too young to remember the horrors of apartheid, but that victory went beyond mere international sportsmanship. The match would later be the focus of the film *Invictus* starring Morgan Freeman and Matt Damon. As Pa hugged me tightly against his chest, I knew that this was a new time, a new country and that this was our game!

"I wish your dad could be with us today," he said sincerely, respecting my biological father and the bond we were building in his absence. I never forgot that day, that amazing feeling, or the love I had for my new rock.

Rugby became my world, the oxygen every young Afrikaner boy breathed. My bedroom walls were covered with photos of the best players. I had no brothers and spent hours by myself in our garden, passing the ball into thin air, running to pick it up and pass it back again and again. I rushed down our greenery, faking out the wings on the outfield, and dove victoriously over the line, imagining the All Blacks behind me in hot pursuit. A tree's branch became my imaginary crossbar, and I kicked the ball for hours on end until I got a shot to go over. Generally, my aim was poor, so I always played positions on the pitch that didn't involve kicking. But speed was the most important quality in sports, and I had plenty of it to spare.

My first game was playing for the Bryanston Learskool's under nine team as a seven-year-old. In the earlier years of South African rugby, the first three grades were bunched together into one team. Bettering the older kids was the only way the youngsters could earn their stripes. I was small, for my age, so I leveraged my quickness and agility at every opportunity. Traditionally, on Saturday mornings, we played barefoot, no matter the weather. In the winter, our toes quickly became numb, and any touch shot sharp stings through our bodies. I learned to regulate corporal pain from an early age. Later in life, I understood the adage: "Rugby is a hooligan's game played by gentlemen." Competing gave me the greatest high. Why do drugs when you have rugby?

My goal in life was to play for my own country, to proudly call myself a Springbok. I was no different from any privileged white youth growing up in post-apartheid South Africa. I didn't recognize skin color or political affiliation. Any Springbok was a god, and I wanted that glory. I threw my life at this dream, trained relentlessly, and pushed myself beyond the breaking point. And then, sure enough, I broke—literally, and all my dreams vanished before my eyes.

My ligaments snapped, and a knee gave out during the 2007 season. I was playing a local club derby match against our archrivals. I was on cross defense chasing down the opposition's fullback. We were both running at full speed, ears back, heads low. As the distance closed, I knew this wouldn't end well, but I couldn't back down. We collided with a sickening thump, which stopped both of us dead in our tracks. The crowd let out a sigh and averted their gaze at the train wreck. The sudden halt in momentum ripped the ball from the fullback's hands. I felt a sudden pain in my knee but couldn't pinpoint exactly from where it came. I saw stars, but I was young and regained my bearings quickly.

"Well done, Alex. You saved us a try there," our captain said as he patted me on the back.

But the next scrum finished me off. The eighth-man picked up the ball from the base of the scrum and broke to the blind side, running straight at me. I bent my knees as I shot low towards him to take out his legs. We hit. However, my left knee disagreed with my momentum. I collapsed on the ground as it painfully dislocated. The sick sound of ligaments ripping was that of whips cracking on a ranch. We lost the game and my season was over in an instant. I underwent surgery to replace my ACL and have the loose bits of cartilage scraped out. I followed up with extensive rehabilitation for a year. Despite my doctor's prognosis, I was hell-bent on getting right back onto the pitch.

The waiting is the hardest part. I followed my therapist's orders with military precision, and in the process, got back into fighting shape. My knee was as strong as ever, and I was ready. I eased back into rugby in a friendly match. My opposing center broke the line and came sprinting towards me. He stepped to the right, and I dove after him. I manhandled his jersey and held on for dear life. I wasn't about to let him get away. As I brought him down, I heard a snap, and a sudden sharp pain shot through my wrist. I immediately knew that something was wrong. In true fashion, I brushed off the injury, taped up my wrist, and finished the game as if nothing had happened. But it was clear that I had torn those ligaments and would need surgery again. My body was showing the accumulated damage, and it never completely healed to one hundred percent. With one injury following another, I had to retire. Life felt unfair, as I was always told that hard work paid off, and that anything was possible. It was now all a big lie. I never competed in rugby again.

I was the third child and only son of a successful businessman. My siblings and I were the wellsprings of his pride and satisfaction. But Dad was unexpectedly diagnosed with colon cancer and passed away a mere three months after his diagnosis, days before my fourth birthday. The most vivid childhood memory was of him lining up his brood of kids every morning before work. He was in polished shoes, a pressed suit, and crisp hat, and made us say in unison: "The de Bruyns vow to always do their best." That was the most uplifting part of my day. But then the morning came when he was no longer at the kitchen table waiting for us. I was too young to comprehend the loss of a parent, but his loving discipline stayed with me forever, and I continued to seek it in other ways.

After his death, the mixed-family Brady Bunch years began. Mother eventually remarried a successful radiologist, who had three daughters from his first marriage. His wife had unfortunately also passed away from cancer. Though I experienced heartbreak early on, I was now settled into a loving family. I felt a great responsibility being the only boy in the mix, and as such, leaned heavily on Pa, who I now loved as much as my late father. Though my siblings had different surnames, we were one undivided clan.

Mother and Pa provided us with a comfortable, upper-middle-class existence. They both encouraged me to think big and freely. No dream was absurd, and my mind lit up with thoughts of what life had in store. I took solace knowing that I had my parents' unconditional love, win or lose. Pa always footed the bill for my many missteps but never complained. "Of course. You're our boy," was always his doting response. He did often wish for me to fail somewhat but only as a means of learning the hard way, and getting back in the saddle.

Since speed and jumping ability go hand-in-hand, one day at primary school I decided to try my favorite Olympic event, the long jump. During a practice session, students leaped from the improvised base of a disused pit, as there was no board. When it was my turn, I came hurtling down the stretch at full speed, my legs scrambling in a blur. My timing was perfect, and I launched myself into the heavens. As I was bringing my legs to the front of my body to land, I saw the entire pit pass beneath me, like the end of the runway after a plane's takeoff. I braced for a hard landing. I crashed on the unyielding, compact grass with full force but miraculously, didn't break a bone. I felt nothing but jubilation at my feat. My coach's face was expressionless, for he couldn't believe what had just happened. I was told never to do that again.

My athletics career blossomed, and I competed against the best athletes in the region. We all pushed each other to excel, but fate dealt me another cruel blow. Just before high school, I severely injured my Achilles tendon. It took another year to heal, and I never regained my springing ability. One of my teammates was a natural born athlete who managed to avoid a major injury. We both made a pact to compete for South Africa. Only he kept up his side of our promise, and in 2008, I watched Robert Oosthuizen compete in Beijing.

Next, after watching one too many X Games video clips, I turned to motocross. I got the "Of course. You're our boy," thumbs-up, and went off into the bush to find dirt ramps to practice on. In short order, I ended up in the hospital, with fifty-six stitches down my side, twenty-five down my leg, and a day of amnesia. When I finally came home, Mother made my motorbike mysteriously vanish. Like a Silicon Valley investor, I didn't fear one failed venture after another, because I knew that I would eventually hit pay dirt.

As the years passed, I warmed to the idea of upholding my family name and heritage. I had to fill the shoes of both a successful businessman and a distinguished doctor. Thus far, every goal I had started ended with me in hospital. I was blessed—or cursed—at simply being "pretty good" at lots of things but exceptional at none. I appreciated how fortunate I was to be growing up in a prosperous family in what was still a developing country. I needed to give back to society and wondered how I would make both my deceased and living father proud.

In 2009, I was at last wrapping up my studies at the prestigious Stellenbosch University, getting an honors degree in Management Accounting. My family was happy that I had left my wild days behind to focus on books. It seemed that I was finally tamed. One day after lectures, I loosened my laces, and lay on my dormitory bed, staring vacantly at the ceiling as the room slowly closed in on me. From the second floor, I could hear my mates playing touch rugby outside on the grass.

My very first existential crisis was triggered by a visiting professor from a neighboring university who was also a Marxist. In his lecture "The Problem with Rich Kids," he asserted that children of wealthy parents have a tougher time defining themselves and quantifying what they'll amount to, for there is unrealistic pressure to succeed. He went on to explain that those kids tended to value personal success more than civility and benevolence. When their achievements did not amount to what they or their parents expected, it resulted in deep anxiety and depression, which in turn reinforced their lack of empathy and decency.

I was about to receive a world-class degree that my parents had paid for, one that I secretly didn't give a shit about. I wallowed in ungrateful self-pity over my genteel future life as an accountant. I wasn't ready to be normal, nor did I fancy polo, or cucumber sandwiches. I wanted to blaze my own radical and extreme path. Four-course dinner party or campfire, I would have the most fascinating story to tell.

I dragged myself out of bed one morning and walked over a minefield of dirty clothes, books, and empty take-out containers. I opened the window and looked out over the residence's muddy rugby pitch. The June air was crisp, and for the first time in a week, the rain had ceased. Feeling nostalgic, I watched the game play out and then spotted Julie crossing the street towards the student center. Little did I know that she was a covert French Foreign Legion recruitment officer. She didn't know it either.

Julie was petite, with curly blond hair, and often wore blue jeans with a white bomber jacket with a furry hoodie. She held her head up high and walked with a purpose. Her toothy smile was radiant and could light up a stadium. We took several of the same courses, and only after a long courtship did I have the nerve to ask her out. But there was a queue of past and aspiring male friends in her life. I locked her in, at least on paper but never felt that she fully gave herself to me. Doubt is a bothersome little worm, especially when I had treated her with the utmost consideration.

"You don't need to change a thing," she always assured me, as if my doubt had a distinct scent.

But one evening, she mentioned that a distant former boyfriend had left to play rugby for a club in France. I forcibly rationalized this as good news, until the other shoe finally dropped. She confessed that she was also going there to visit him. I politely ended the conversation by saying: "Julie, don't get on that plane." But all the tears she cried didn't keep her from doing exactly that.

I was humiliated and felt less of a man. I was unable to play the sport that I loved and lost a girl to a rugby player with a real career. If I hadn't blown my knee, would I have Julie around my arm? How could I win her back?

Weeks later, after a night of drinking, I heard somebody pounding on my door. I opened it and came face to face with Julie. She wanted desperately to have me back. Her remorse and contrition almost moved me to tears.

"I shouldn't have gone to France…" she pleaded. I ran my fingers lightly through her hair, smiled warmly, and cleared my throat.

"But it's too late. Get your big girl pants on, harden the fuck up, and leave!" We always hurt the ones we love. As soon as she left, I broke down like a baby.

"No, you, Alex, harden the fuck up," I said to the mirror as I slapped my face. "'The de Bruyns vow to always do their best.' You don't need anyone. Prove it to the world…"

The next morning I washed the salty tears from my cheeks and ventured downstairs and outside to join my mates Jannie and Benno in the *braai* grilling area.

"Alex, where you been, bro! The fire's almost ready for you to cook our meat," Benno shouted from a camping chair, beer in hand.

"I thought you ladies would have done all the *braaing* by now?" I said, forcing myself to lighten up. Jannie got up gingerly to give me a warm handshake.

"Good to see you, Al. Shouldn't you be preparing for your exams?"

"In a few months' time, I may never see you sorry *okes* again. I gotta live for the moment." The truth was that I no longer cared whether I passed or failed. "Hey, where'd Benno go?"

"He's taking a leak," said Jannie, nodding towards the bushes.

"Dude," I exclaimed, "the coals will be cold by the time you finish putting on your make-up!" Benno, a burly athlete, casually flicked an empty beer can, hitting me squarely on the forehead. We had a good laugh out of it, and I slowly felt like myself again. We cracked open more lagers, watched our boerewors cook, and talked rubbish until the conversation turned to our plans after graduation.

"I told my dad that I want to be a life-long student," Benno said, trying to fool us into thinking he was a serious academic. Jannie smiled and shook his head. Benno was an exceptional rugby player who was already in talks with a provincial club. I was happy for him, as he deserved every bit of his success, and yet, I couldn't suppress the deadly sin of envy.

"Alex, what are your plans?" Jannie asked, breaking my train of thought. He looked at me intently and with sincerity. He was a few years our junior, but we took him under our wings. I admired his cool and quiet confidence. Growing up with sisters and having lost my father, I longed for the companionship of a brother. Jannie became the one that I never had.

"Yeah," I responded nervously, fishing for a good answer, "you know, Pa has a few business ventures. I'm sure he'll put me in charge of something."

21

"Looks like it's the French Foreign Legion for you," Jannie retorted with an impish smile. "That's the only option for a curious and masochistic guy with balls like you."

"The French what?" Benno interrupted, as he bit into the pork chop skewered on his fork.

"Don't you remember Sinatra's old song: 'Bet they'd welcome me, with open arms. Do we march together down the aisle? Or do I march that desert sand...are you gonna be mine or *au revoir cheri*? It's the French Foreign Legion for me!'"

"So the Frogs put all the pissed-off men of the world into an army to battle Arabs in the Sahara?" I asked.

"Yeah, what better way to rid Paris of pesky foreigners than to use them to fight wars without endangering actual Frenchies? Oldest trick in the book," Benno mumbled with his mouth full. "Like that old Gary Cooper movie *Beau Geste*. To protect their family's honor, three rich English *okes*—brothers—take the blame for a stolen diamond. They fuck off to the Foreign Legion in Algeria and are given a new name, a new life, and a new start. Only one made it back alive..."

"My uncle was a Selous Scout volunteer," Jannie added "He did some awful things in Rhodesia. This part of Africa was a very different place in the seventies. Alex, you were born a few decades too late. But you could probably get into the British army. Unlike the South African forces today, they're still fighting in a few parts of the world."

"Our countrymen were slaughtered by the Brits in the Boer war," I said. "Fuck the Queen's Army. If there's a country that keeps order in Africa, it's France."

"I like how you think, Alex," continued Jannie, who knew me as a brother, and could read my mind. "You want experiences and adventure over material possessions."

"You guys know anybody who joined the Legion?" I asked.

"Yeah, man," Jannie said, "my cousin had a friend whose dad enlisted just before he was born. Bloke never talks about it, though."

"Did he finish his contract?"

"I heard he barely lasted two months of training. It's that tough. Then again, nobody's ever met someone who actually served, and there are lots of bullshitters. I don't even know if they take South Africans, but you can try..."

"Sounds romantic but no thanks. I've got nothing to run away from," I said defensively. "My family worked hard to give me what I have. I got lots that I should be running *towards*."

Benno stared at Jannie and burst out laughing. "Relax! We were joking about the Foreign Legion. You wouldn't last a day. You march to the beat of your own drum. The only contract you're going to sign is with Ernst & Young. Your five-year commitment will be marrying a hot Afrikaans woman and starting a family."

Jannie was more somber. "Alex, my big brother, I have a few more years here. But wherever you end up, know that I'm pulling for you. Don't forget about me..."

Over the next few weeks and between exams, curiosity got the better of me, and I researched the Legion more intimately. I was fascinated by the idea of sacrificing one's life for another country. What son would die for those who didn't believe in what he was doing?

Recruits pledged allegiance not to a foreign state but to the Legion itself. A French passport awaited those who survived the first five-year contract. Aside from being a haven for men who'd given up on life, it was also the most highly-trained army in the world. On a moment's notice, they were deployed to the nastiest parts of the globe. The Legion marched for days on end and was expendable. It consisted of the world's best prior-servicemen and only accepted ten-percent of applicants. Women were barred.

Founded by royal decree in 1831 and headquartered in Algeria until the 1960s, the Legion was solely used to protect and expand the French colonial empire. Though initially banned from stepping foot in Europe, the Legion participated in every major French operation, from the Carlist Wars in Spain, to both World Wars. The Cinco de Mayo celebration in North America commemorates Mexico finally kicking the Foreign Legion out of Puebla. It made up the primary contingent of French forces fighting and dying in Indochina before the Americans took over.

The Legion's most recognizable white headdress, the *képi blanc*, embodied its Saharan origins. Olive *képi* covers that were originally issued to new recruits were eventually bleached by the sun, demarcating one as an old hand.

In modern times, the Legion fought in the Persian Gulf and sent units to Afghanistan. It ultimately evolved into a rapid deployment force to preserve unsavory French interests in the deepest reaches of Africa, from Chad to the Congo, to Djibouti. Outside the purview of international law, the Legion operated as a renegade army, eerily similar to Mad Mike and his merry men of yesteryear. Unsuspecting legionaries were shipped to the Amazon, French Guiana, to police the frontier against drug smugglers and protect the European Space Agency's primary spaceport.

With a new burning curiosity for the Legion, I suddenly snapped and decided to join. I'd been viewing my life through the lens of an anonymous soldier redeeming himself through suffering. I would finally become a hero—a legionnaire—or die trying! The next few months were a blur, mostly sweat but many tears as well. I now had a motivating fire within me. Although my degree became an afterthought, I hit the books hard, making sure I passed my exams. "I checked off that box for you, dear dad," I said pointing to the heavens. I wanted nothing to hold me back from my new dream.

I was afraid of springing this surprise on my family, so eased them into the idea by announcing that I wished to spend some time in France after graduation. Pa hugged me and muttered the usual: "Of course. You're my boy." But he suspected something was off, and to allay Mother's fears, offered me a job working for him. The family was now pressuring me, and there was no way I could let them down. I accepted his offer, but my heart was elsewhere. I used that time to save money, rehabilitate my wrist and knees, and get into the shape of my life. In due course, though I still couldn't bend my wrist, it was strong enough for me to crank out a few pushups on my fists.

When I got to the point that I couldn't continue faking my work duties, I announced that I would join the French Foreign Legion. From the expression upon my parents' and siblings' faces, nobody was convinced that I would be back any time soon.

"Come on," I pleaded, "why the sad looks? Listen, whatever happens, I'm going to make you all proud. I'll return in a few years with a chest full of medals from missions all over the world."

Goodbyes were said, tears were shared, high-fives swapped, and I placed a wad of cash in my pocket. I was pumped and primed. Yet, all the while, something said to me: "Alex, don't get on that plane…"

Looking out of the window, I didn't want to miss a second of my beloved continent passing below me, from the tip of the Cape to the Atlas Mountains of North Africa. In nearly every expanse that passed, legionnaires lay buried in unmarked graves. As we approached Europe, the South African Airways Boeing dipped beneath the clouds, allowing me to see a vast expanse of deep green water, the Etang de Berre, spread out beneath us. Beyond this lagoon, a speck in the distance was the Mediterranean city of Marseille.

I grabbed my backpack from the luggage collection area and followed the SORTIE signs to the exit. Standing nervously in line at Passport Control, I watched as a disinterested clerk interrogated a young woman in front of me. I made up a thousand reasons for coming on a one-way ticket from Africa. If I admitted that I'd come to kill the enemies of France, would I be given the red carpet, or deported on the next flight out?

The clerk merely flipped through my passport after scanning it, stamped it, and blurted "Next." I now actually wanted to announce to all that I was joining the Legion. But how many French civilians truly cared about foreigners dying to protect their freedom? We were simply cannon fodder for dirty wars.

I stepped out of the airport and breathed in the crisp air, the fresh scent of a new start. I quickly made my way outside and could barely get myself a taxi, a quick lesson on how difficult life was going to be without knowing a lick of French.

I was dropped off near the muggy main port. I walked past Fort Saint-Jean, the medieval keep that until very recently, was used to corral would-be Legion volunteers heading for training in Algeria. As if taken from *The Count of Monte Cristo*, recruits slept on stone floors atop a dirty layer of hay. Armed guards loaded them onto transport ships—many perished en route. Luckily, conditions for volunteers had now improved somewhat.

Evening approached, and I sat at a terrace bistro for a drink of milky Ricard Pastis. I found a cheap hotel nearby and enjoyed my last night of peaceful rest. I primed my bed and emptied my pockets onto the nightstand. My hand trembled as I held my passport, prepared never to see it again. There was no turning back now. "This one's for you, dad," I said. "I know you're with me every step of the way." From my opened wood shutters, I could make out the street sign at the nearest intersection: Légion Etrangère Recrutement. I faintly heard: "Of course. You're my boy."

THE ELECT

Each Legionnaire is your brother in arms whatever his nationality, his race or his religion might be. You show him the same close solidarity that links the members of the same family.

Code d'honneur

I tossed and turned all night and woke that morning well before my alarm. I lay still, staring blankly at the moldy ceiling of my old tawdry room. The light had penetrated the shutters, and like an alcoholic on a bender, I didn't know if it was dusk or dawn. I heard Renault horns and police sirens in the distance as France's second largest city also rose from its slumber. I felt completely hung-over, though I hadn't drunk a drop. My anxiety came from fear. This was my second day locked in my room, buying myself time, as cabin fever enveloped me. I was inching closer to doing what, up to now, was merely a fantasy, a cruel joke over grilled boerewors.

As a means of reassuring myself, I flipped through the crumpled Foreign Legion pamphlet that was still damp from armpit sweat. "*Esprit de Corps. Discipline. Action. Eficacité.*" I had some cash, cold feet, and was already homesick. A thousand excuses for not joining raced through my mind. Benno's last words lingered in my ear: "Don't be like those lesser men who gave up." But how would I feel if I came home with my tail between my legs, having chickened out at the last minute? I'd eventually have to confess the truth to my lager-swilling mates over grilled meat.

But just as I lapsed into sleep again, a screeching sound next to my ear shocked me out of bed immediately. I threw my cheap alarm clock across the room and started packing my meager belongings into my rucksack. I grabbed it, my wallet, and passport and headed for the door. As I reached out for the handle, my hand trembled. Cold beads of sweat formed on my forehead, as I prepared to face the school ground bully, knowing that I was outmatched.

"I'm ready for what may come!" I said as I left the comfort of civilization for the last time.

I strode towards the Gare de Marseille-Saint-Charles as the icy gusts pushed me along. I wasn't prepared for a European winter and hoped that the Legion had better cold-weather gear than I did. I had the false impression that France's Mediterranean coast was always hot and steamy, replete with sunbathers year round. If I'd done some proper research, I would have seen that the Legion's training center was tucked inland, before the foothills of the Pyrenees.

With the very last of my Euros, I bought a regional train ticket to Aubagne, outside of Marseille, headquarters of the French Foreign Legion, the 1er *Régiment Étrangèr* (1er RE). I took an empty seat next to a window, wanting to enjoy the last melancholy moments of Southern France as a tourist. The sound of the train's wheels chugging over the tracks lulled me into a trance. I imagined myself bunkering down in some flaming town, brothers-in-arms to my side shouting *"Cover me!"* as we ran from one stone hovel to another. We'd push forward, like a well-oiled machine, to capture the enemy without losing any comrades.

The sudden whistle of the train approaching my stop woke me back to reality. I instinctively reached for my wallet as I considered the possibility that the Legion wouldn't accept me, in which case, I'd be hungry and homeless in a foreign country. But I subconsciously wanted to have no material belongings to go back to, not even a coin to make a phone call.

I walked off the platform and approached a stranger who was leaning against the wall finishing his cheap Gauloises cigarette.

"*Pardon*. I'm looking for *la Legion*."

"*Oui, oui*," he said in a thick accent. "I am going that way if you want a lift."

I accepted, as he flicked his cigarette and opened the back door of his old Citroen. We pulled off at a speed I didn't think possible for such a jalopy.

"You wish to join our *Legion, oui*? I often see many sad men drifting up the hill to do the same."

"Damn straight," I smiled. He flew around a traffic circle while searching for his lighter, gesticulating wildly.

"I live here, and I don't know what on earth happens in that regiment *sacré*. I've always wanted to be a legionnaire," he said, lighting another smoke, "but I don't feel tough enough. In another life, without kids and a wife, I would try. You are lucky because have nothing to lose. We all have goals, but how many of us pursue them? Life passes fast, and you make your choices. Dreams and nightmares, sometimes they are the same thing, *eh*?"

The Frenchman interrupted his own monologue and stopped the car next to a stone wall. On it was a flaming grenade military insignia besides the words LÉGION ÉTRANGÈRE. "This is the end of the line," he said, pointing a bony finger at what lay ahead for me. He lit a third cigarette before speeding off again, with a sympathetic, regretful frown on his face.

I now stood before the bricks and mortar embodiment of my life's hopes and desires. Still in a trance, I scanned the red-roofed building in front of me. I passed through the low stone wall via a narrow gate that read ENTRÉE, and then followed a few steps towards what looked like an information window. This is where I finally had to pull the trigger.

My heart beat through my chest. My legs became numb and weak from adrenaline. My steps slowed, unconsciously, as if to give me time to change my mind. All I could hear was my inner voice saying: *Turn around. Go home. You're unfit. You'll fail. This really isn't what you want.* I felt this way when my parents took me to church, as I often felt unworthy to be in God's presence. But it wasn't the Almighty planting seeds of doubt now. It was me, and the voice was now a scream: *They don't want you! You're not brave enough! Make up an excuse as to why you weren't accepted! Nobody will know!* But I would know...

I knocked on the glass window, straining to see inside, and then glanced at my watch—12h05. Nothing happened. I knocked again, louder. I then heard rushed footsteps approaching from inside and the door swung open.

"*Savez-vous pas qu'il est temps de déjeuner, idiot!*" a burly red-faced *caporal-chef* (chief-corporal) shouted at me in irritation, pointing to a green steel bench along the wall. I barely made out that he was calling me an idiot for bothering him during lunch. He slammed the door shut, and I sat down and waited. As expected, my first encounter with the Legion was a proper ass-chewing. At least it wasn't an ass-beating. An hour later, from out of nowhere, a bawling Romanian in tattered clothing fell at my feet.

"I came from my country to join the Legion! Now I have no money, no hope…" he fretted.

"Come back with papers!" a different *caporal-chef* said as he slammed the door behind us again.

13h30 and still no-one let us in. 14h00 and I started to wonder whether I should try once more. The medieval rule of Saint Benedict stated that a novice should be rejected three times by the monastery before being accepted into the order. At 15h00 I was about to knock again when the door opened. The first *caporal-chef* appeared, wiping crumbs from his mouth and smelling of vinegary wine. How dare I disturb his meal? I must have been asking too much by volunteering to die for his army. Then again, I was bad at dealing with pride. I always expected to be thanked for a favor. The Legion would change that attitude quickly.

"*Donnez-moi votre passeport!*" he snapped, generously dousing my jacket in spittle. I nervously handed it to him.

He turned and walked away with purpose. Not knowing what I was supposed to do, I followed him through a green gate that lead into the belly of the beast. In front of us was a large parade ground, which I recognized from old dusty books and photos. In the center of it stood the famous *Monument aux Morts* that was dragged to France from Algeria when the Legion left their original home in 1962. Every stone was hewn from Saharan granite, carved out and polished by generations of legionnaires. I could picture the pioneer Sappers in leather aprons marching in unison, with axes over their shoulders, their beards immaculately kept and their *képi blancs* glowing.

"*Dépêche toi, putain de merde!*" my escort yelled, snapping me out of my daze. I feared doing or saying something wrong. Like a dog, I slowly caught on to the tone of the French speaker's voice and what he was pointing at.

We walked passed a courtyard on our left, where there were several dozen recruits in blue tracksuits running sprints over a twenty-meter cemented area. Near the fence, there were a few pull-up bars, and just beyond that, a tall structure with three thick ropes hanging from it. Several combat-uniformed men were shouting at the *bleus*—as they were known—getting their egotistical jollies. But those soldiers didn't display any actual rank and seemed equally clueless. I got the odd feeling of watching show dogs in a pen being forced to perform for the pleasure of their handlers. It all smacked of power and sadism. I suspected that I would join the "general population" sporting those blue tracksuits in a few days after initial processing.

I was rushed to the next building beyond the courtyard, into a bare reception and waiting area. It felt like my own South African home affairs office where I had queued all day to renew my passport before coming to France. The *caporal-chef* called me to the counter, and a copy of the enlistment contract was shoved in my face. "Sign now," he said. "As of this moment, you are in the French Foreign Legion." *Lord, what did I just do?*

Everything was signed by hand and stamped by an official. Hardened legionnaires spent entire days shuffling bits of paper with India ink. A working computer was nowhere to be seen. But this was *de riguer* ever since the Legion opened its doors to men who were best never heard from again. I was about to become but an inkblot on an officer's ledger. From dust we come and to dust we shall return.

Over the course of the afternoon, two other recruits trickled into the room. The first was a young lad with dark eyes and sharp facial features. He wore a black polo neck jumper and a grey leather jacket. A dragonfly was tattooed on the side of his shaven head. He must have been in an early morning punch-up with the police after a night of drugging and clubbing, as his face was banged up. Without making eye contact, he sat down on the opposite end of my bench, sliding forward into a slouch, captivated by whatever was on the ancient cathode ray television in the corner. He wasn't interested in talking, so I turned to the recruit who sat down behind me.

"Marcos," he introduced himself with a deep rich voice.

His friendly eyes burst forth from under his monobrow, giving warmth to his round face, which produced a smile the moment I reached to shake his hand. His grip was firm, something I always respected in man. Marcos had a dark complexion, jet black hair and didn't seem a day over thirty. He was six-feet-tall, with strong, broad shoulders, an approachable demeanor, and had the whiff of a Fortune 500 company CEO. For some reason, I immediately liked and trusted him.

"I'm Alex," I said. "You speak English?"

"Yes. Portuguese, Spanish, and French too."

"Where on Earth are you from?"

"Brazil. I grew up in Columbia, and have been studying French for two years preparing for the Legion."

By nightfall, we were now a dozen blokes in the small room. Marcos, the unofficial *responsable*, greeted each one with his classic handshake and an endearing mixture of their language and his. He was just the sort of guy who made anybody feel ten-feet tall. He was also a regional Brazilian ju-jitsu champion but didn't boast of his accomplishments. Unlike me, he seemed absolutely primed for what lay ahead.

"Have you any idea what happens next?" I asked.

"It's Friday, and apparently they only take in new recruits on Mondays. So you're stuck with Mr. Personality over there and me for the whole weekend."

I never got the memo about recruitment taking place only on business days. Real men don't read the fine print. So like a young Siddhartha, I learned to wait calmly, and with my passport confiscated, that was all we could do. Every single volunteer that was thrown into our drunk-tank looked like he was carrying the world on his shoulder. "I bet this guy's a former cop," I said.

"No, club bouncer in Warsaw," replied Marcos.

Everyone was broken as if life had been kicking his ass, and this place was his only way to tap-out. Old Legion memoirs dispelled the myth of it being a refuge for criminals and scumbags. The face of the underclass, as Orwell put it, was simply that of the hardworking, downtrodden man.

"Jakov over there is on the run from the Hungarian authorities. He killed a guy in a bar fight last week and faces life if he's arrested," Marcos told me as he shared the story behind every cheerless mug. Jakov sat timidly silent the entire time, afraid to even hurt a fly.

"Do you believe him?" I asked as the young man didn't look like he'd ever been in a fight in his whole life.

"Judge not, lest thee be judged," Marcos reminded me, as we tried to make sense of the boasting, the bullshit, and the truth. "You never been to jail? Everybody's innocent; even the bloke caught on video. You'll learn quickly that every man has a past, as well as a few lies that he himself believes…"

Eventually, the boredom brought out bizarre or unpleasant behaviors. Guys would pace back-and-forth relentlessly from one wall to the other. One recruit spent the day counting every single brick that made up an adjacent classroom. Dragonfly became physically aggressive and began shoulder-checking others and then cursing at them to provoke a fistfight. Another guy bumped his head on a table for two hours straight. *That's a good way to keep your sanity.* I thought sarcastically.

Finally, a corporal came into the room crisply shouting: *"Douche. Dormir."* Not understanding a single word, I followed in Marcos' footsteps to avoid getting smacked.

We all then stripped down and queued for the showers. Feeling rather uncomfortable, I finished as quickly as possible. Twenty bunk beds were precisely arranged ten per side, one meter apart. Each was covered in a cheap brown polyester blanket that had probably never been washed. They would have lit a city block if viewed under a black light. Hygiene aside, I was happy to have something soft to crash on and tried to get some well-deserved rest.

I lay awake most of the night, trying to block out the sounds only a dormitory full of strangers could provide. Gasps from the back sounded like a man drowning in his own saliva. Then the farts began, mixed with the aromatic essence of feet and armpits. Bedsprings creaked as everyone tossed about. Every hour, one unfortunate alcoholic got up and paced the room. The anxiety from being confined by the Legion's walls was getting to him. Up and down he went, between the beds, to the door, and back. Alcoholism is a disease of the night. I kept one eye open.

As I was barely dipping into REM sleep, the lights came on at 04h00. But suddenly there was a commotion in the back of the dormitory. Jakov and Dragonfly were up in each other's faces. Marcos and another ran up and pulled them apart.

"I saw you going through my wallet, you Russian piece of shit!"

There was a nasty zero-tolerance policy for thieving in the Legion. Since we all feared getting sent home for fighting, Dragonfly got off easy.

"That bastard would've been lynched if we were in a regiment," Marcos confirmed.

"It doesn't really matter. If you're accepted into the penitentiary on the other side, the Legion confiscates or throws away everything on your person," I mumbled as we got on with our day.

"*Petit-déjeuner, dix minutes!*" we heard behind us as a smartly dressed *caporal-chef* came through the door. Everyone scrambled to get dressed and ready to follow him for our first meal inside the Legion.

We chaotically trailed our fearless leader and joined a long line of recruits waiting to be served. Not knowing what fare to expect, I was both nervous and excited. I hadn't eaten in a day and could have instantly wolfed down a full English breakfast—toast, fried eggs sunny-side up, bacon, and baked beans on the side. But I was rudely corrected.

When I got to the front of the line, I noticed everyone picking up a bowl. I presumed we'd be getting porridge with a thick pat of butter. However, the bowls were for nothing but coffee. "Is this it?" I accidentally asked out loud as I snatched an accompanying croissant. I then felt a blow against the back of my head.

"*Tais-toi, kurwa!*" the supervising corporal shouted, and then gave me a second smack just to remind me that I was at the bottom of the food chain in these hallowed halls.

After finishing breakfast, which took me an entire five seconds, we were escorted back to our holding room and left to our own devices. Some guys killed time doing pushups, sit-ups, or dips with the help of two chairs. It was the time-honored way of establishing authority and who was the alpha. A few scrawny black Africans were running about whom I was concerned for, especially when the Russians scowled and spat in their direction.

By now I'd almost memorized every word in the recruitment videos played one after another. Every few hours a new recruit would be tossed in. We'd all size him up in seconds as if we were in a no-rules fight club waiting for the bell. But even with the simmering discord and cultural animosity, there was some male bonding. We all came from different backgrounds and joined for various reasons, but we were now in the same shit together. Misery loves company.

On Monday morning we were finally escorted to a medical hall to start our entrance tests. By now there were fifteen of us, most of whom had arrived that day–they must have read the memo in more detail. I had my blood tested and was otherwise prodded and poked to confirm that I was healthy enough to be fired at. I was then ushered into a tiny interview room. I found myself sitting in front of an on-loan French army psychiatrist who looked like the male version of Nurse Ratched from *One Flew Over the Cuckoo's Nest*. I wondered what kind of psychological games we were about to play to tell whether I was sane enough, or too sane, to join the Legion. He wore a wrinkled uniform and had his name pinned on his breast: Bernheim.

"Do you have a criminal record?" he asked, not looking at me. Was that a trick question—would a rap sheet put me in front of the queue?

"No."

"… petty crimes?" he continued, appearing to be writing down his observations.

"No."

"Ever killed anyone?"

"No, but I'd like to right about now," I muttered under my breath.

He finally looked up at me, peering over his small round spectacles. Without a word, he closed my file, set it aside, and grabbed the succeeding one. "Gentile soldiers were born to fight. *Bon*, fit for duty." He nodded his head towards the door. "Next."

The subsequent step in the process, known as the interview with the Gestapo, brought me before a sergeant within the French intelligence service, the *Deuxième Bureau* of the Foreign Legion. This is where every man was stripped bare to his soul. While the Legion would protect a man's identity at all costs, they hated secrets.

The English-speaking *sergent-chef* interrogator leaned forward on his elbow. He was wearing the formal mud-green jacket of the Legion with several insignia and badges on his shoulders and chest. His face was wrinkled by years in the sun, and he moved like an arthritic pensioner, though he didn't look that old.

"Name," he asked.

"Alex de Bruyn."

"Sir—I'm sure you mean to say, *sir*."

"Yes, sir. Alex de Bruyn, sir."

"Why do you wish to join the French Foreign Legion?"

"I want to be an elite warrior. I'm looking for adventure, and I never back down from a fight, sir."

He looked me in the eye. "Young man, do you have an education?"

"Yes, sir."

A corporal then suddenly entered the room, grabbed me by the lapels, and accused me of lying. After shaking me about, he was then told to wait outside.

"Sorry for the corporal. He means no harm. I'm not like him. I just want you to be honest. Now, do you have a family?"

"Yes, sir."

"On this sheet, write down your life's story, each name, place, date, school, job, bank account, girlfriend. Leave nothing out. Oh, and write only on one side of the page."

My fingers cramped as I recalled every minuscule detail.

"Fine," he said barely skimming my scribblings. He then paused and weighed me up. I held his gaze. He then relaxed, sat back in his chair, and folded his hands in his lap. "You will surely desert within two years," he shrugged, smugly. "You have too much going for you in your real life."

Unperturbed, I spoke confidently. "I will prove you wrong, sir, and if selected, I'll finish first in my class. Consider it written."

Pride swelled up in me. Nobody tells a de Bruyn that they'll fail.

He leaned forward again. He must have seen many a man come and go through this office. "Promise me one thing—the day you realize you have had enough, pack your bags and quickly get the hell out of here. Understand?"

I bowed up. "That won't be necessary, sir,"

Rage filled my soul, displacing any lingering self-doubts. Was this the *sergent-chef*'s way of mind-fucking me? I was determined to be a Five-Year Johnny, come hell or high water.

"You are a proud young man. I am a proud old legionnaire. But I also recognize that there is a certain pride that kills." I sat motionless without even blinking. "As you wish, lad," he added with a slight smile on his face, breaking the awkward silence. "So, if someone comes to Legion and asks for you, what do we tell them?"

I gave a vague response. "I have nothing to hide, sir."

"As of now on your name is DARKLAY, Albert," he said as he signed a piece of paper and closed my file. "You are dismissed, but remember, if we ever meet again in two years after you desert, I will not be so kind. Consider this our last talk as gentlemen."

Albert Darklay. My *nom de guerre* sounded something out of a comic book, spy thriller or *Donnie Darko* film. A man without a country, I no longer belonged to either my homeland or my family name. My soul was on loan to the Legion. I felt a mixture of excitement and apprehension at the thought of trading all that to play soldier. But I knew that in just a few years, I'd have an epic story to tell my grandchildren.

An hour later, Marcos and I were decked out in blue tracksuits and cheap canvas sneakers. Carrying a plastic bag of soap, disposable razors, and toilet paper, we were then chucked into the courtyard under the afternoon sun with several hundred other buoyant men hoping to pass the next stage of selection.

By hook or by crook, each morning all *bleus* were taken on a five-kilometer run, most likely to spot any unfit rejects. Upon return, most of the *bleus* were assigned work detail. It wasn't by accident that there was not a single rose petal or Gauloises butt to be seen on any Legion parade ground. Since the 1er RE had hundreds of *bleus*, it was the shiniest regiment of all. Before a volunteer got his hands on a rifle, he had to master the intricacies of a shit-clogged toilet with a coat hanger. Nobody spoke as they worked. And as there was no common language, we communicated by grunts and threats.

Those not on work-duty hardly had it better since they did nothing more than stand around the prison courtyard, complain, and hate their lives—compared to navel-gazing, washing a mountain of pots and pans was a treat. There was pushing and shoving, with the bigger blokes having their way with the weaker ones. The bullies all seemed to speak Russian.

One morning Marcos was tasked with de-weeding the flowerbeds near the mess hall entrance, and I got kitchen duty. By chance, I ran into the only other Anglophone in the pen, an American former Marine. He introduced himself by his *nom de guerre*, Daniels. The Legion generally saw American recruits as soft but made a special exception for the "rebels" from the former Confederacy. Daniels had attained the rank of sergeant and survived two tours in Iraq. His bright eyes were not yet dulled by war, but they held a tinge of sadness.

"What interested you in the Legion?" I asked as this seemed a good opener to any conversation while we were both elbows deep in milky dishwater. He turned his head and exposed a dragon tattoo across his neck. "After I got this stupid thing, the Corps wouldn't let me reenlist," Daniels answered. "The Legion became my next best option."

"Mate, now how about you tell me the real reason you left the Marines?" I kept digging.

"I'm from a long line of Southern fighters. My old man joined the Corps when he was sixteen. My granddad was also a Marine. But I was fucked up after my second stint in the sandbox. I tried the civilian thing, Civvy Street, as you guys call it."

"And now you want to try the military life again?" I continued.

"My mind wasn't right, and still ain't. I wasn't a good civilian. I burned my bridges with the Marines. So I came here to work through my shit. I'll take the devil that I know. Pick your poison."

Speaking with his shoulders back and head erect, Daniels was a born soldier, a man who knew nothing but how to do that and do it well.

"So what happened on your second tour?" I asked trying not to badger him.

43

"It wasn't *that* stint that broke me but the time after it. I've seen Death, and he actually has an appealing face. Fighting is addictive. When I got home, I couldn't handle the silence. I'd have nightmares of friends being blown up by roadside bombs, or Iraqis found in torture houses. I'd wake up in a cold sweat with birds chirping and kids playing outside. The VA shrink put me on medication, but they only made shit worse. I hated those smiling civilians in their SUVs who had no idea what we endured for their pathetic asses."

"Weren't most of you guys well received after the war?" I asked naively.

"Ain't no such thing as a real-life hero. Once you're back in civvies, nobody gives a shit. But the journalists love us. We give them some damn good stories. I now see soldiering as just another job. I don't need all this Red White and Blue, Stars and Stripes propaganda. Do you know what PTSD is?" he asked.

"Um, the bad effects of fear and stress?"

"No. It's your conscience telling you that you're a piece of shit. You ever been in a situation where you could have saved a life but didn't because you were scared? You ever let a man burn to death? You ever did anything that haunts you for the rest of your life? We ain't saints or sinners. We're just the unlucky bastards tested by God by being thrown into ethical situations that would make a civilian's skin crawl."

"How do you handle your PTSD?"

"By being in combat. That's the only time I feel normal. I think the Legion will offer me lots of that. If it doesn't, I hope to die in a blaze of glory—with my boots on. I'm already dead," Daniels finished.

Every man had his story.

Seeking safety in numbers, *bleus* clustered in different *mafias*—groups based upon language and nationality. The most prominent was the *mafia russe* who occupied the north section of the rear yard. They spent much of their time lifting heavy objects and intimidating anyone walking past.

Our tiny *mafia anglaise* consisted of any native English speakers and Germanics. We occupied an area on the gravel overlooking the smashed up basketball court. Our favorite pastime was talking shit about the filthy Russians. But unlike most other groups, we had decent lives to return to. We even had a few thinkers and romantics among us. With our fine grasp of the English language, education, and straight teeth, we were sometimes an arrogant lot. As such, Anglophones were hated by the Legion as a whole, and especially by the *mafia francophone*.

Down on the court, any unfortunate not part of a *mafia* wandered around alone in circles for hours, trying not to get on anybody's bad side. This prison yard was where everyone lost their minds, and *instruction*—basic training—was where some found a semblance of sanity.

Amid the *bleus*, there were those who passed selection and were officially slotted for *instruction*. This gave the elect a small jump in rank amongst the volunteers and the title of *rouge*. They shed their tracksuits and proudly wore Legion-issue combats. Plato said that the measure of a man is what he does with power. Thus, being promoted to *rouge* was too much of an ego boost for many. Feeling that they were already corporals, the *rouge* immediately bossed around the *bleus*. Nature abhors a vacuum, and this led to violence. But the *bleus* had to suffer in silence since they too aspired to be *rouge* and not get tossed out for standing up for themselves.

Sure enough, we witnessed this dynamic when two diminutive and likable Senegalese chaps got into a massive punch-up over something as petty as bed inspections. Two days earlier they were sitting, both in their blue tracksuits, laughing and sharing food that they'd smuggled out of the mess hall. The smaller of the two was promoted to *rouge* first and was on call that night. They calmly argued in French, but things quickly got heated. And then out of nowhere, the *bleu* cracked the *rouge* right in the jaw, sending him stumbling backward, holding his mouth as if he was pushing his teeth back in.

"*Putain de merde!*" the *rouge* shouted as he launched himself towards his aggressor, tackling him straight into the corner of the bed.

The pair spent the next five minutes rolling on the floor like women in a mud-wrestling match, each trying to get in a cheap shot. It wasn't an impressive scrap but a fight to the death nonetheless. Either bloke would have gouged out the other's eyes if given a chance. Brawls in the Legion are "Two men enter; one man leaves."

"*Garde-à-vous!*" the burly Polish duty corporal shouted as he entered with a look of disgust on his face. Everyone sprang to attention, except for the pugilists who were still rolling about.

As if he had caught two dogs fighting, the corporal kicked the larger of the two, who was now on top, right in the ribs, which separated them immediately. They were marched out of the gates never to be seen again. The abuse and resentment of power cost these two men their life's dream. Then again, maybe God smiled upon those poor sods and got them out of a shittier situation than that from which they had come *Peace be with them.*

Every evening, the compound was called to the assembly block to stand at attention before the corporals. The colder it was, the longer we would wait, with any movement resulting in pushups. "Who wants to leave?" we were always asked. One need simply raise his hand, and he'd be escorted to freedom with no questions asked. It was a shameful way to go, but I did envy them. Only decades before, *bleus* were held under lock and key. But these days the Legion only wanted men who wished to be there—unless, as I learned, they could force you to stay.

Many of the volunteers I walked in with eventually tapped out, including Dragonfly. I wasted hours questioning my own decision to join, wondering if it would have been more exciting to sit behind a desk as an accountant. It took a superhuman effort to stay, but I convinced myself that this was just part of the bullshit I had to endure before the proper soldiering began.

The latter stages of selection consisted of a physical fitness test and a psychometric one to assess both one's personality and reasoning skills. But to become *rouge*, we had to pass every examination just to make sure a volunteer wasn't a complete waste of oxygen. Not surprisingly, there were scores of wealthy Latin American kids who had just graduated from private schools. In the mix were also drug addicts, homeless men, and French teens that'd gotten upset with their parents that morning and decided to show them who was boss.

Since its inception, the Legion was renowned for accomplishing superhuman feats of physical fitness. The Legion could march for weeks on end with little or no food, with several legionnaires expected to die along the way. To even be considered for selection, one had to be an exceptionally fit man. But as the Legion was dragged, kicking and screaming, in line with the regular French army, times had changed. Yet, like all laws, organizations always find a work-around. And while my generation of recruits may not have been supermen, the Legion would make them into such in due course.

The physical standard was to perform a mere three pushups and four pull-ups, with a VO_2 max test to measure endurance. Even then, I was astonished to see how many failed that basic requirement. "Your preparations getting here won't be wasted," Marcos said to me, noting my disenchantment.

"But I thought the Legion was the best of the best?"

"Look at all these assholes in the courtyard trying to get into this tiny army of eight-thousand. They'll only take a handful. The garbage gets taken out early, and the cowards ask to go home. But if a guy's still here, like us, then he's a damn good prospect. The cream always rises to the top. Don't be put off by the stupid formalities. Sweat saves blood, and the unprepared will curse the day they were born after the first march."

The Legion only selected roughly thirty recruits per month to be shipped to the southwestern town of Castelnaudary to begin *instruction*. As the days passed, only Marcos, the Yankee and I were left from our initial rag-tag bunch. Fate rewarded tolerance and diligence. As the entire parade ground stood at attention one morning, the *cadres* called up Marcos and me. I'd stuffed sachets of sugar in my pockets after lunch and was terrified that I'd be called out, beaten in front of the others, and sent home. But then a dozen other names were announced. All I could understand was "*Section rouge...Castelnaudary ...instruction.*" We made it!

We were all then marched off together to the parade square, saluted the tricolor, and led to the material stores, free from harassment from our handlers. In line with all tribal rites of passage, we too participated in the Legion tradition of shaving each other's heads and crowning ourselves with the coveted Green Beret.

"You like my round, pale, shaved head?" I asked Marcos as we high-fived.

Uniformity and conformity is a strong male essence. Once all of us looked exactly alike, akin to bald, tattooed convicts, we were issued our combat kit. Women love dresses and shoes. Men love sturdy uniforms. Like kids on Christmas morning, everyone bowed up proudly as we donned our new combats. The feel and color of the denim were crisp, as was the smell of our fresh leather boots. One could have put a broom handle in our hands, and every one of us would have sworn that we were trained killers. Boys can be a simple lot, and I was no different, but the Legion knew this all too well.

After ragging and goading each other over our new gear, we got into our high-quality, immaculate dress uniforms and were led to a very special corner of the regiment. We entered the *Musée de la Legion* through steel gates shaped in the Legion's flaming grenade, past various battle flags enclosed in glass. In the cordoned off section of their own crypt, on the walls, etched in gold, were the names of each officer who had sacrificed his life. The Foreign Legion, modern Catholic crusaders, celebrate death. Any soldier can fight—the legionnaire can die. The blood of these fallen volunteers washed away our pasts as well. The Church Militant and Church Triumphant—those in hell don't count—a church doesn't simply count the living as its current members. Though we were nominal Christians, at best, along with a few Muslims and African animists, the Legion stood firm with its politically incorrect Christian traditions and heritage. If one expected special meals, holy days, or specific chaplains, then this wasn't the army for him. Nobody knew what to expect until we were formed around a neon-lit cross embedded in the granite floor. I felt chills down my spine as I stepped over it to get to the other side. "Christ knows your sacrifices," I whispered.

Standing shoulder-to-shoulder along the walls of the adjacent room, we were handed our signed five-year contract by a major who had come to welcome his new recruits.

"*Engagé volontaire* Darklay," the major announced as he stopped in front of me, his uniform pristine and the golden lines on his epaulets glistening.

"*Présent Majeur*," I proclaimed feeling ten-feet tall, as I immediately came to attention and raised my right, open palm, to salute our demigod. Afterward, we filed reverentially past the encased wooden hand of the heroic *Capitaine* Danjou. The experience left me breathless. I belonged. We all belonged.

After being dismissed, fantasies were shared of what we hoped *instruction* would be like. Guys who knew someone who served retold stories of how tough it really was. No matter which books one read or what story we heard, nobody could fathom what to expect. I was about to find my own truth, even if it cost me my life. *Walk the walk, and if you judge, do so only after marching in another man's boots.* I was shedding my old skin, not knowing that sometimes, the old wine will rupture fresh new wineskins.

Within an hour after being chucked back into the holding-pen, every *rouge* forgot that moments earlier, they were still *bleus*. *Camarades* became slaves in an instant. Instructions were barked out as soon as the corporals left. I refrained from abusing the others but was helpless at keeping the bad apples from doing so.

The remaining week of toil and boredom was perfectly bearable, as the camouflage combats versus the blue tracksuits made all the difference in life. If we could remain *rouge* and never actually be real legionnaires, that would have suited most of us.

A week later, our sugar high came to an abrupt halt when our *instruction* corporals reared their ugly heads. We waved to the desperate looking *bleus* and lugged our kit onto a French army bus that would take us to our awaiting train.

Waiting on the platform were two impeccably dressed sergeants who wore black *képis*—reserved for NCOs and above. They were flanked by two equally smart looking corporals wearing *képi blancs*. Although sergeants in most armies were responsible for training level tasks, the Legion's glacial promotion regime meant that corporals were in charge of our daily duties, while the *cadres* provided mostly organizational leadership. Officers were an even rarer and princely lot. With the odd exception, they were French citizens and army academy graduates who were only concerned about finishing their *secondment* to the Legion and future political careers. Many Algeria-era veterans spoke fondly of Captain Jean Marie Le Pen, former leader of the far-right *Front National* party.

Caporal Bordon was a muscular, olive-skinned Argentinean specimen of a legionnaire. His Anglo surname didn't seem to match him, and he was rumored to speak fluent English. *Caporal* Basil, our other instructor, was a short, stocky Bulgarian with a kind face.

"It's going to be awesome," Marcos whispered as we walked towards the gate, a full military issue bag in each hand.

"*Qu'est-ce que c'est?*" *Caporal* Bordon shouted at Marcos, who seemed a bit too proud of himself.

"*Rien, carporal!*"

As Marcos let out the last syllable, Bordon gave him a massive right hook to the gut. Marcos dropped both his bags and fell to his knees, trying to recoup his breath.

"Nothing will be awesome," *Caporal* Bordon muttered as he stared me down.

"*Oui, caporal!*" I responded, helping Marcos to his feet, realizing that life would now be a little different than that in Aubagne.

"Welcome to the Foreign Legion, faggots!"

I rocked from side to side as the train left Aubagne and chugged through the picturesque countryside of Southwest France. I was pensive, looking down at my fingers, playing absentmindedly with my new identity card. Something didn't feel right.

There were over thirty other recruits in the car, about a third of them francophones, a third Russians and Eastern Europeans, and a third other Western Europeans, Latin Americans, and a few Asians. Though the Legion was restricted to non-French foreigners, there was a loophole for native criminals that allowed them to enlist as Belgians, Swiss or Canadians. Among us was a diminutive but cheerful Mongolian. Nobody else spoke his tongue, so he got by on the little Russian that he knew. He barely seemed old enough to be in the Legion, and we had no idea how he had passed selection. But he was generous, honest, and hardworking—too saintly for a place like this. The *mafia russe* nicknamed him Myshkin. Also in the mix was a serious looking central Asian of an unknown origin who no one dared look at or touch. He kept to himself and stayed under the radar. Harrak was a Moroccan francophone, petty drug dealer and a practicing Muslim who had somehow convinced himself to join the remnants of France's colonial forces.

I looked forward to getting my hands on assault rifles, automatic pistols, grenades, and rocket launchers. I imagined long, grueling marches in the snow of the Pyrenees to the song *Eye of the Tiger*.

"Darklay!" shouted *Caporal* Basil. "Back here, now!"

I expected to be belted for helping Marcos up off the deck.

"*Oui, caporal,*" I said trying to catch my breath.

"I hear that you play rugby."

"Um, I do, or did. I don't really compete anymore."

"You know that the Legion has no such sport. Watch yourself..."

But then he cracked a warm smile.

Escaping the Amazon

"Hey, so tell me, did you play for South Africa? The Springboks are my favorite team! I played rugby in my home country, but it's still a small sport. I'm not so big, but I was good. Really, I was."

It was obvious that he was a far more skilled super fan than he was a player, but I was pleasantly surprised by his demeanor.

"And don't even get me started on the All Blacks World Cup Final of '95. Listen," he finished in a low tone, "we'll talk later. Just hang in there during *instruction*."

By the time the train pulled into our destination, heading directly into the setting sun, we were all alive with enthusiasm. But the cozy atmosphere of our railcar was contrasted by the bone-chilling gust of air coming down from the surrounding mountains. Winter was still in full force. I realized that I should have skipped all my months of preparation and joined in the summer. I'd have been in the regiment by now.

Just as we arrived, our salty looking sergeants then announced that we would be heading to the *Quartier Capitaine Danjou* in Castelnaudary, the 4e *Régiment Étrangère* (4e RE). From that day on, simply hearing the word "Castel" muttered sent shivers down my spine. We then hopped onto the waiting Legion coach. It lurched forward and slowly made its way up the winding country road…our road to Damascus.

IRON SHARPENS IRON

> *Respectful of the Legion's traditions, honoring your superiors, discipline, and comradeship are your strength, courage and loyalty your virtues.*
>
> Code d'honneur

It was the perfect season to see Castlenaudary in the territory of Lauragais. The outlying fringes were spread out and nestled amongst majestic trees. The setting sun glistened on the Canal du Midi, highlighting the riverbank, which burst with a blend of ancient and modern buildings, stacked closely together in layers of centuries past. Several church spires pierced the skyline. *How could this pictorial town ever be our house of pain for the next four months?*

We passed the old Legion training center, tucked inside the city, to the sparsely populated outskirts near the railroad tracks. Eventually, the road curved into the entrance to *Quartier Capitaine Danjou*. As we rolled through the wide gate, I saw a large maroon parade square directly in front of us, looking just like the pictures I'd seen, with perfect rectangular lines of legionnaires standing in formation. I misted up at the thought of now being part of that.

We piled out of the bus and were led to our barracks. They were basic but impeccably tidy—Mother would have been proud—with rows of beds, aligned with the floor tiles. Cream lockers stood between each bunk. Down the passage was the communal ablution area where we were given five minutes to piss or shit before reconvening. "Back down the passage, and then turn right down another corridor, left, and it is the first door on the right," the *Caporal* Basil rattled off in indecipherable French.

All thirty of us were then squeezed side by side on small wooden chairs in a tiny lecture room. A pissed off *Caporal* Bordon materialized and marched front and center. He wore long-sleeved camouflage combats with a black patch on his left shoulder adorned with two green stripes. On his head was the brilliant *képi blanc*. If God willed it, we might wear that same coveted headdress in a month.

"*Bienvenue à l'instruction*," he started in French. "You are now in the second phase of becoming a legionnaire. Your asses belong to the Legion. Your allegiance is to the Legion, and you no longer exist. We don't give a shit about who you were in your past life, nor do we care about what sins you have committed. We're not interested in which military you served or your previous rank. You are all now equally worthless."

He pulled out a pair of glasses from his left chest pocket and picked up a clipboard from which to read, switching into pamphlet propaganda mode. "You are no longer *rouges* but *engagé volontaires*. Tomorrow you will go to a farm where you will spend a month adapting to the Legion lifestyle. There are no modern comforts there. You will master the French Army issue assault rifle, the NATO-compliant FAMAS-F1. We will teach you how to march. You will learn French and how to sing. We will show you how to fold your clothes. We will teach you how and when to eat. You will learn how to set up a bivouac, everything you need to survive. Do not assume you know anything until *we* have taught it to you. If you successfully complete a month on the Farm, you will qualify for the *Képi Blanc* march. If you complete this, you receive your *képi* and continue three more months of training here at the 4e RE."

The dreaded *Képi Blanc* march was lore in the Legion and sent shivers down the spine of any aspiring recruit or veteran. It was as mysterious and gossiped about as sex to prepubescent boys. It would make or break the *section* but was the only thing standing between scumbags like us and that funny, little white cap.

He looked at us and spoke off the script.

"Other military units brag about the difficulty of their Hell Week. This is your Hell Month. If you finish *instruction*, you will apply for your regiment of choice. The five best recruits may choose their placements, the rest of you—tough shit. Before summer, your ass will be on a flight to serve somewhere in the world. Enjoy France for now. You may not see it for five years."

He mechanically placed his spectacles back into his pocket and marched out of the room.

I am really and truly in the military now, I thought, as fear and adrenaline coursed through my veins. The next morning our troop transport rolled through kilometers of dry bush, interspersed with patches of old snow before turning onto a small dirt path leading to a rusty gate. We disembarked and marched two more kilometers up a steep embankment to what would be our home for the following month, to the famed *Bel Air la ferme*, the notorious Legion shack for beating street rabble into fighting men. Behind a two-story stone building was a barn. Two large front doors opened to a terrace. Surrounding it all was deceptively perfect lawn—there was no question as to who would be manicuring every inch of it.

"All we're missing are the cows," I joked to Marcos who was fumbling with his gear and panting after the trek.

To the right, there was an overgrown field with soccer goals on it, which was unforeseen for an elite military training facility. Yet by now, the Legion had conditioned me to expect the unexpected. Behind the main building ran a narrow staircase to the second floor. We were soon herded up those stairs, like lambs to the slaughter, by shouting corporals.

"*Dépêche toi, kurwa!*" was the standard command. It seemed that one could get by in the Legion with knowing a dozen curse words, many of them Slavic.

"*Trouver un lit et descendre en dix secondes!*"

Scrambling about, I had no idea what was commanded, except that I had ten seconds in which to do it. In our sleeping quarters, I dumped my bags onto one of the top bunks. The steel frame rattled as they landed on the crusty mattress.

"You'll sleep like a baby," Marcos quipped, as he took the bottom bunk. "Just don't piss the bed."

After everyone staked his territory, we were ordered back down the stairs in single file and gathered outside on the main lawn before the barn. We stood in perfect rows, bowed up, feet together, arms at our sides, looking straight ahead at *Caporal* Bordon in front of us.

"You learned nothing while you were in Aubagne, and look like bags of cellulite. Fifty pushups!" he barked.

We all dropped to the floor.

"Not here, ladies. Around back!" he continued, pointing to a square patch of dirty snow behind the barn. We ran over and started again.

"*Un! deux! trois!*" he counted loudly. Every time someone didn't drop low enough, he commenced from the beginning. Just to mix things up, he repeatedly ordered us to hover in mid-position. "*Au milleux!*"

Caporal Bordon's training philosophy was basic. Why use a carrot when the stick works so exceptionally well? We would become accustomed to performing pushups in the snow. Never mind our feet, our palms were in danger of getting frostbite. "You only need to know one thing to leave *Bel Air la ferme* as a legionnaire," he began as we approached fifty. He slowly walked back and forth before us, deliberately creating drama to accent his Al Capone speech. "Loyalty and obedience are virtues. A legionnaire is a fighting man, not a thinking man. As long as you put in an honest effort, the Farm will be easy. After all, I am a fair and good-natured person."

"*Oui, caporal!*" we all mumbled in unison, as we shifted our weight. It wasn't our muscles that were screaming for relief but our damned hands. My mind drifted blissfully to my days at University when I was the hall-chief instructing the new first years to do ten pushups in the quad. *How the wheels turn.*

"*Son…bitch,*" I let slip.

"*Qui était-ce?*" *Caporal* Bordon queried, disgusted that someone had muttered a word.

Silence. His footsteps quickly approached. I felt Bordon's eyes penetrate the back of my skull. His feet stopped squarely by my side; his boots immaculately polished with a fresh smell radiating from them.

Thump! My lungs collapsed with the blow of hard rubber and boot leather against my solar plexus, leaving me gasping for air. Like a fish out of water, I flailed on the snow trying to fill my chest before I suffocated. As my face turned purple, *Caporal* Bordon bent down and gripped my collar. I momentarily thought he was executing an act of mercy, but instead, he comforted me with an iron fist to the gut. This may have been intentional since it opened up my lungs and kept me from soiling my pants.

"Listen carefully, Darklay, you white South African piece of shit," he said in perfect Queen's English. "In Argentina and France, you Anglophones think you are better than us. I have a Scottish grandfather, but I'm fucking proud of my brown skin. I'm Latin and Catholic. This is your last warning. You're on my shit list."

"*Oui, caporal,*" I managed to get out through my gasping and coughing, "but, you see, I'm not really English. The Dutch were in South Africa before—"

"Shut the fuck up, *gringo de mierda!*" he screamed and followed with another blow.

Attention all passengers: After an unplanned stop in France, we have just crossed into Crazytown.

We knocked out three hundred more pushups before being ordered to stand at attention again. "And do them right next time!"

We were dismissed to wash-up for dinner. We were starving and salivated at the thought of getting something in our stomachs. As we stood around our tables, empty mess tray in hand, *Caporal* Borden then confirmed what was on the menu.

"Enjoy your cup of coffee, you bastards! Do exactly as ordered and you might get some milk and sugar next time. *Bon appétit.*"

"*Bon appétit, Caporal!*" was the standard response. Borden then left to dine in the adjacent room with the *cadres* on spaghetti carbonara, specially prepared by an *engagé volontaire* who was formerly a chef.

"Where're the damned toilets?" I whispered to Marcos. "I'm about to burst."

"You didn't smell them while we were doing pushups? Check out that rusty shack over there."

I walked back over the dirty snow to the outhouse behind the barn and opened the creaking door. On the ground was a white ceramic slab sloped down into a brown hole in the center. Covered in muddy boot prints were two rough patches on the slab. I situated myself on them and made sure that I was aiming properly. As I heard the corporals shouting, I pushed as hard as I could and ended up nearly slipping on my excrement in my rush to join the others.

There was never a moment to breathe. We were completely self-sufficient, from preparing our own meager meals, washing our uniforms, shoveling snow and patching leaky roofs—everything is done *pas gymnastique*, double-time. The Legion was deliberately stressing us, forcing us to think quick on our feet, to manage chaos, and to perform under tremendous stress—practice as you play. Before any task, we were called to stand to attention in front of the barn. More often than not, *Caporal* Bordon found an *engagé volontaire* with a whisker out of place, and we were punished with punches or ever more pushups.

Right or wrong, there was a particular Legion way of doing everything, and any deviation or improvement was met with a swift penalty. Like a failed state, the Legion was structured in a way that those with power stayed in power by abusing that power. Early on, it was easy to write off any abuse as being simply part of training. Coupling that with our ignorance of French made for a toxic witches brew of resentment. But it did motivate even the dumber recruits to learn fast.

French lessons occurred daily, right at the time when we were most fatigued. The corporals were keenly aware that there would be plenty of men nodding off, fresh specimens for sadistic gratification. I was soon the unfortunate who became that night's whipping boy. We'd already been jerked around all afternoon with running hills, pushups, and other wild goose chases. My eyelids weren't just heavy, but I'd developed a splitting headache. *Caporal* Bordon's words slowly melted into opium-induced gibberish.

"Jump in for me," I heard Julie say, as I found myself standing on an iceberg.

"Um, ok. But why am I in the arctic?"

I jumped, and then suddenly snapped out of my trance as Bordon dumped a bucket of ice water over my head. I let out a gasp, as my heart skipped a beat and my lungs contracted. Everybody around me tried to keep a straight face, for laughing would result in the same punishment for them.

"Now clean up your mess!"

As imagined, what was meant to be a French lesson descended into both a circus and a good reason to scrub every inch of the farmhouse. With enough pushups and punches, even a monkey could learn a foreign tongue.

But I was surprised to discover that I had a knack for languages, as knowing both Afrikaans and English gave me an advantage. By the end of the first month, I was virtually fluent in "Legion French," which was understood by street thugs and immigrants but indecipherable to local first graders. Our parlance had always been influenced by the times and campaigns of the Legion. After two world wars, Germans had a disproportionate influence on the Legion. With the fall of the Soviet empire, the Legion was largely a Slavic army. While we still used words like *achtung* (German, warning), the more commonly heard word was now *kurwa* (Polish, bitch). Like salt and pepper, it was used to spice up every other sentence and usually added at the end.

The Farm was a place far from the public's eye where corporals could use any method to transform us into a synchronized marching and fighting machine. As the Legion washed away our sins, in return we had to abandon our bad habits. Our *cadres'* careers and reputations were at stake, and they had to present a perfectly squared-away product once we returned to Castelnaudary. The process was the ancient technique of breaking a man down and then building him back up. The problem was that the "building back up" stage was skipped over. Every light at the end of the tunnel turned out to be an oncoming train. This was the Legion's finishing school, not for upper-class gentility but for the unwashed proletariat.

I stubbornly clung to the idea that if I were a good soldier, the corporals would reward my effort. The Legion followed the Japanese corporate maxim: "The nail that sticks out will be hammered." High performers were punished, or simply given a heavier load. I strategically hid in the herd, seeking safety in numbers. I saved my best performance for when it counted. When it was measured, I kicked ass.

The greatest challenge was not always the corporals or the brutality of military life but living with a dozen different nationalities pressed into one *section*. Cultural animosities that never existed before suddenly exploded. The "diversity is our strength" mantra thrown around in politics, and popular society was proving to be utter nonsense. Indeed, in *Mein Kampf*, Hitler wrote about how, growing up, he knew little about Jews and had in fact, never met one. What kept nationalities in the Legion from killing each other wasn't cultural tolerance but a force-fed, totalitarian Legion culture. Spartan techniques were used to enforce cohesion, at the risk of eating our own youth. If an *engagé volontaire* made a mistake, the *section* paid for his sins. While peer discipline was effective if used sparingly, in the Legion it created a lynch-mob frenzy over the slightest infraction.

One blowup occurred while we were being punished after Daniels spoke out of turn. He was ordered to watch the *section* perform burpees. As they approached thirty, Harrak muttered something about America deserving the attacks on 9-11. For a good reason, the Legion forbade any political discussion among the ranks, but some were bound to learn why the hard way. Daniels was a seasoned brawler after earning his stripes in the nastiest roadhouses in North Carolina. Before any of us knew what was happening, we heard the distinctive thud of bone smashing into bone, sounding like a 2x4 hitting a tree. Daniels had landed a high Maui Thai roundhouse kick to Harrak's head, which immediately dropped him like a sack of potatoes.

Caporal Bordon rushed in and broke up the commotion. Harrak was revived and found a new urge to go another round with the Yankee. As such, Bordon made it easier for the combatants to finish what they had started by following a classic Legion tradition of ordering them to dig a trench in the rocky ground, where they could brawl to their heart's content, with only one man making it out. The mob of *engagé volontaires* wanted to see a good cockfight. The creepy Central Asian was stretching his limbs as if preparing to then take on the winner, but by the time the pit was finally dug, Harrak and Daniels were best *camarades*. In the back of my mind, I managed to give the bastard *Caporal* Bordon a shred of credit. He may have predicted the peaceful resolution from the start, and this was his way of engineering closure and allowing the fighters to save face. Unresolved disputes tended to fester and die a slow death. Violence was a tool, and the Legion was a master craftsman. *Caporal* Bordon winked at me.

The Legion was always known as a light and extremely nimble unit. Even in recent conflicts, the Legion refused to equip their men with body armor because they considered it too heavy and bulky. The Legion perfected the art of improvising and making do with what was available. As such the lightweight 5.56mm FAMAS (*Fusil d'Assaut de la Manufacture d'Armes de Saint-Étienne*) fit the Legion's unique mission.

It was nicknamed *le clairon* (the bugle), because of its stumpy bullpup shape. It's configured with the ammunition feed behind the trigger. The receiver housing is made of a steel alloy and the rifle furniture of fiberglass. A relatively long barrel for an assault rifle makes it extremely accurate. It's designed to withstand water, mud, sand, and extreme temperatures. Its trigger guard can be pulled away for firing with gloves in arctic conditions. It can fire single or three-round bursts, spitting 1100 rounds per minute in fully automatic mode. A flip-up feature allows it to launch grenades from the barrel. But for all its worthy features, it has a reputation for jamming. Cleaning every small and highly engineered part was a nightmare and took several hours. We then had to memorize the names of the thirty intricate parts in French. Our onsite armory was guarded by a corporal 24 hours per day. Yet even with such strict protocols, there were cases of *engagé volontaires* who managed to desert with their FAMAS, the most serious sin in the Legion. Authorities were reportedly authorized to shoot them on sight.

Our morning runs through the pristine countryside were getting progressively more exhausting. Even the most unfit *engagé volontaire* was essentially competing in a 10K race every day of the week. Yet the treks were only a means of conditioning us for the more important objective: the long march.

True to form, the Legion inducted us into this beastly tradition early on. Each march was gradually longer and more grueling. They were meant to prepare us for our final judgment, but our bodies barely had time to recover before the next march was announced.

After an hour of marching, heads dropped, and backs bent low to spread the heavy load as the straps cut into our shoulders. My arms were stabbed by pins and needles from the lack of blood circulation. Our legs and backs were strained to breaking point, and even a spoken word felt like a waste of energy. A slight touch was an extra burden. One followed behind the bloke in front, every man for himself. If anyone took a step to the right or left out of marching train, a slew of Slavic curses would be hurled at him.

"*Dix minute de pause!*" were the sweetest words ever barked.

My *section* passed out by the wayside, helpless. I pulled off my rucksack and threw it on the ground with great relief, joyful at getting rid of the unbearable load for a few moments. Never in my life had a time of rest seemed so luscious and revitalizing as when I lay stretched out on the French snow. To my surprise, the other men kept their rucksacks on to get in a few more seconds of rest and save the effort of lugging it back on. During the repose, blood collected in our limbs, which made standing up and walking again excruciating. Our worn-out legs protested at the previous and upcoming abuse. We looked like a crowd of senior citizens slowly wandering down the road. "The devil marches with us" was a common Legion maxim.

General Oscar de Négrier, the legendary nineteenth century commander, was known for the famous quote: "Legionnaires, you became soldiers in order to die, and I'm taking you to a place where you can die!" He was beloved by the Legion and considered any wounded man a hero. But when he saw an exhausted legionnaire stumble out of the ranks and collapse during the awful marches in Madagascar, he showed no mercy. He muttered the three words that have defined the Legion: "March or die!"

During the conquest of North Africa, treks, which no European commander would attempt, were commonplace. One covered six hundred kilometers in sixteen days, with men fed on nothing but rice. In the Sahara, a legionnaire who fainted on the march was tied to the baggage cart. A pole was pushed through the sides of the wagon at the height of a man's arms, and the legionnaire was roped to it by his shoulders. The pole kept him in a standing position—the cart rolled on. He either had to march or was dragged along. Legionnaires were horrified seeing this torture but afterward understood why it was necessary. The fighting value of the Legion depended on its marching ability. If a legionnaire were to become separated from the company in the desert, he was a dead man walking. Hostile Arab women, who were far crueler than the men, soon surrounded the helpless man, who suffered a horrifying death after being horribly mutilated. While the Legion was proud of its marching tradition, it came at a great human expense. We modern legionnaires, trekking across France had it relatively easy.

After marching through the entire afternoon, we usually set up a bivouac late at night. Coming from warm South Africa, it wasn't long before the novelty of snow wore off. The Brazilians and Vietnamese felt the same way, though the Russians were used to it. We slept in snow that went up to our shoulders—if we managed to get any rest at all. Depending upon the time and altitude, temperatures flitted between above and below freezing, which meant that we were not only frosty but wet, adding insult to injury. We cursed and shivered, but the cold wasn't about to go away, so most of us put up with it—for now.

Though we thought the day would never come, it was now time to conquer the *Képi Blanc* march. On the eve prior, we were allowed to pack our own gear. Marcos and I carefully selected our essentials.

I crammed all my food into the bottom of my rucksack. Two tins of duck liver pâté, two packs of crackers, three sachets of oats, and powdered milk. I left the tinned mutton and dehydrated vegetables behind and opted for one of the few pleasures on the Farm—the Legion marched on Mini Babybel cheese.

"Why are you leaving food behind?" Marcos asked.

"I'm going light. By the second day, even a tin of sardines will feel as heavy as a brick."

I inspected my cheap cotton sleeping bag, and it was so worn that it looked more like a bed sheet.

"How cold do you think it'll be?" Marcos asked, trying to stuff a polar undershirt into an over-packed rucksack, hoping not to hear my expected response.

"Mate, we'll be walking in the Pyrenees, and it's dead winter."

"Shit, I can't take the cold anymore," he said in a trembling voice. I sensed that this was more serious than the garden variety bitching.

"Get yourself together. We can do this. If you don't grab my balls, we can double up in my bivouac."

Early the next morning, dressed in winter combats, we were dropped off at the foothills of the Pyrenees with our *section* leaders. With FAMAS slung over our shoulders and rucksacks secured, we followed the lieutenant through the knolls towards the gentle looking but sinister white slopes ahead. We didn't follow a particular path but simply headed west to the Spanish frontier. Like sirens, the mountains rose up towards the bright blue sky, irresistibly pulling me towards them.

The foothills were broad and grassy with an occasional rock protruding from the earth. We walked along a stream for most of the morning, hopping from stone to stone. Most of the recruits were still in a jovial mood, with energy to spare. Even at this age, ignorance was bliss. However, the terrain became steeper and rockier the higher we climbed.

Luckily the sky was clear, and the midday sun warmed us up nicely, but the weight of our full kit soon became unbearable as we continued the steep climb in one go. The chatter and laughter petered out and was replaced with heavy breathing. Two hours in, we were now gasping. The air was thinning, and men began falling behind. As punishment for not "motivating" the laggards, we were deprived of the upcoming rest.

"My first blister just popped. I can't make it," Marcos whimpered as he held on my rucksack to avoid stumbling back down the steep incline.

"The devil isn't marching with you. I am," I assured him, though I had no energy to spare beyond mere words. For a split second, I wished Marcos, my best *camarade*, would simply die—pain and fatigue were fucking with my mind.

By mid-afternoon, after we had walked uphill for eight hours, a normal workday for civilians, we were finally permitted a brief pause. I gulped down my water and took off my boots to investigate the throbbing that had begun on the back of my heel three hours earlier. My feet were now raw and bleeding, but before I could even consider disinfecting my wounds, we were ordered to get moving again. I put my sticky sock back on, hoping that the pain from my pack and lead-weight FAMAS would mask the pain in my feet.

"Hang in there," said *Caporal* Basil as he walked next to me. "After *instruction*, we'll have a Kronenbourg and watch the upcoming Tri-Nations Cup. By the way, keep an eye on your *camarade* Marcos."

Escaping the Amazon

We arrived at a particularly complicated section that required the use of all fours to haul ourselves over the rock face. It created a bottleneck in the group and gave me a short moment to catch my breath and look around. The streams we had left behind earlier in the day were now just a thin sliver of water, glistening in the setting sun. We were surrounded by a vast, spectacular expanse of mountains that extended as far as the eye could see.

"These same peaks kept the Islamic hordes out of Christian Europe," Daniels muttered to me. "Twelve hundred years later and we're still fighting them."

"I thought that was the Battle of Tours?" I replied.

"Oh, so now you're some kind of Ph.D. historian?"

We finally reached a plateau of sorts, and the terrain was approximately flat. I assumed we were on Easy Street until I felt a solitary snowflake hit my face. Within half an hour, the ground was covered in snow. The mesa was cradled on either side by two steep peaks that loomed over us and blocked most of the sun, which now hovered low on the horizon. We were given one minute to get our headlamps on and then kept walking, barely able to see the man in front of us. I was horrified that I might accidentally twist an ankle, walk off a ledge, or trigger a landslide. It was another three hours of marching before the corporals finally allowed us to set up a bivouac on a snowy crag high in the Pyrenees.

I was exhausted, and my feet were steak tartare. My muscles cramped and my fingers and nose were completely frozen. Now the dropping temperatures would be freezing our sweat. There was only a short period where we were not humid and overheating, and not shivering from the biting cold. Those minutes passed, and my clothes became icy and encrusted. But I loved it, in a masochistic way, like an abused wife claims to love her husband. Pain reminded me that I was alive.

Though freezing and exhausted, before we slept a wink, we had to master the latest winter field craft technique. With frozen sausages that used to be our fingers, Marcos and I somehow began erecting our shared bivouac. We found two trees three meters apart, secured a cord between them, and draped my groundsheet over it to form a basic A-frame shelter. We piled some snow and sand on the corners to pin them down like an ordinary tent. We then placed Marcos's sheet on the ground inside, hoping it would defend us from the cold, or at least any dampness from below. We were the first to finish, and *Caporal* Bordon gave us the evening off from harassment. He sensed that now wasn't the best time to fuck the men around.

I caught my breath and, under a full moon, marveled at my minor feat of mountaineering. Mallory would have been proud. But then the unthinkable happened—a loaded truck pulled into our encampment without any effort or circumstance. I never imagined there to be even a footpath for miles, much less a road. My sense of accomplishment dropped like a stone. Steam arose from the back of the vehicle. The entire *section*, shivering in their cotton sleeping bags, got to see the officers and NCOs enjoy a nice, hot cooked meal, replete with coffee and hot chocolate. But I was far too exhausted to care—sub-zero temperatures and all, I slept like a baby on the snowy ground.

The next morning we woke up in a low cloud, which swallowed up the previous day's majestic views. The ground was covered in a blanket of thick snow. We had five minutes to pack our bivouac and scoff down some liver pâté before continuing our march higher up into the mountains. The water in my water canteen had frozen solid. It was a stupid *jeune* (rookie) mistake to have left it away from my body for any stretch of time. In combat conditions, it could have cost me my life. Sure enough, *Caporal* Bordon went by every man in formation and jostled his canteen to see if any other knucklehead had committed the same infraction. I was the unlucky bastard with a fat lip that morning. I estimated that it would be several grueling hours before I could drink a mouthful of water.

A fierce gust blew wet air from all angles, which penetrated even the tightest seams. My boots, socks, and uniform were drenched within an hour. Marching in the fog was more dangerous than trekking in pitch darkness since lamps couldn't illuminate our path. For six hours we walked with no break—though it was probably to simply keep us from freezing to death. Nature was playing games on me and giving me every reason to give up and succumb to the elements. Had my *camarades* not been motivating me—more like pushing, cursing, and threatening me—I may have given up. By lunchtime, the clouds finally lifted and slowly exposed stunning nameless peaks that were close enough to touch.

"Breakfast!" *Caporal* Bordon barked, which prompted a unified sigh of relief, as everyone could get a moment's rest and a few sugar calories. My mouth was dry, but I still managed to pour a sachet of milk powder down my throat, which congealed like glue. Luckily there was now a gulp of water in my canteen, which I used to wash down my sustenance. I grimaced with pain as the condition of my feet deteriorated.

"*J'ai fini. J'ai fini!*" I heard moaned behind me. The unthinkable had happened. Our best man and alpha Marcos was throwing in the towel. I turned around to see him kicking his rucksack down the hill we were about to descend. "Fuck all of this shit. Fuck the *képi blanc!*" he shouted as he continued to throw the toys out of his crib.

This is not going to end well, and this is definitely not the time to pull this shit, I thought to myself and quickly tried to catch up to him before *Caporal* Bordon did. I was too late.

"Fine, go fuck yourself down at the bottom of the hill with the rest of your shit," Bordon responded with a boot to his back. Marcos flew face first down the mount right into a tree, barely managing to keep all his teeth. "Politicians in Paris won't allow us to execute deserters anymore. You're lucky to be in today's Legion, where every *engagé volontaire* has rights, and we have to show restraint. We made this shit easy, and this is how you pay us back? If we were in Algeria like the old days...you all disgust me."

The *section* stood in petrified silence.

"Your break has just been cut short," Bordon bellowed. "Get your asses moving, and thank *camarade* Marcos for that one. He doesn't need us and wants to get off this mountain on his own. Now repeat after me: '*Merci camarade!*'"

Little Myshkin and I ran down the hill and helped Marcos up.

"Dude, not yet," I whispered to him as I got his kit squared away. "At least wait till we get off this damned mountain, and then have a good think."

Like a rock star's tour manager, by whatever means, I just needed to get my client through the next show. If I could convince Marcos to stick it out one more afternoon, we could give him extra rest in order to gather the strength to finish the march.

"I won't make promises I can't keep," he responded.

74

"Don't think for a minute that this piece of shit *Caporal* Bordon won't leave you here to die," I said in earnest.

"I'm only moving for you, Darklay. But may God forgive me for what I'll eventually do," he said as his eyes teared up.

"Deal," I declared immediately. "You can't be caught out here with your FAMAS. I'm going to slowly take it from you. Okay?" Myshkin, without protest, took his rucksack. I had to find a way to distract Marcos until the evening. For now, we had to quickly catch the rest of our *section*, already a kilometer ahead of us. We continued walking through the snow for another ten hours before stumbling into a small flat clearing where we could set up a bivouac.

"Marcos, help me put up our camp. After that we can get some hot food inside us before we go on sentry duty," I snapped to Marcos in hopes that he'd forget all about jumping ship.

"I made my decision," he mumbled as he got up and walked towards *Caporal* Bordon.

Sitting a few meters away, I couldn't make out what was transpiring between the two, but *Caporal* Bordon remained calm, for he was more interested in setting up his own bivouac for the night.

A minute passed, and Marcos sheepishly returned.

"And?" I asked.

"I was expecting to have my teeth kicked in, yet he couldn't even look me in the eye. He told me that I was free to pack my shit and start walking home. I wish he'd beaten me. Regardless of who came on top, this would've been settled. He called me a coward. His words hurt more than any blow."

"Damn. Now the Legion's fucking you around," I said as I set up. "French law allows you to leave within six months of joining. They're reneging and making it impossible for you to do so. This place is one big lie. *Quid est veritas...*"

The next morning we awoke early to pack up our bivouac. This was the last day of our march, and it was crisp and clear. With a heavy rucksack, swollen knees, and blistered feet, getting off the blasted mountain was as beastly as getting up it. For thirteen hours, we descended from the rocky upper regions, past the tree line, and into the greener foothills. I lost track of how far or how high we had walked, but before I knew it, the march ended abruptly! Most didn't even know that we had now earned our *képi*. Like much in the Legion, this event was largely anticlimactic. I wanted to keep marching, to endure more pain and feel that I merited the award. The *engagé volontaires* with shredded feet or lost toenails did not share my sentiment.

After surviving that first right of passage, we spent an entire day sprucing up the farm and ourselves for the formal ceremony. The daily violence was replaced by tedious training in drill and staging. We perfected marching in formation, arms presentation, salutations, and generally just not being ham-handed thugs. Wearing new specially issued and pressed camouflage combats; we were taken to a distant medieval citadel, which was unlikely to be on any tourist guidebook. It may have been owned by the Legion and used for special occasions, but it smacked of the Welesburg castle where Nazi SS officers were formed and revered as demigods. It was surrounded by high stonewalls that enclosed a sizeable stately courtyard. Knighthood awaited us.

I polished and buffed my boots one last time. In my right hand, inside a clear protective plastic bag was my *képi blanc*, which nobody dared place on his head before the moment we were officially allowed to do so. Though much in the Legion thus far had been ad-hoc and half-assed, the Frogs did take pomp and ceremony seriously. Despite its foreign character and brutality, the Legion beatified the best of French culture in its feminine reverence for beauty and presentation.

The Legion takes arrogant pride in being the only force that marches to the slow 88 steps per minute, compared to the standard 120 of nearly every other military unit in the world. The "we don't hurry for anybody" attitude is on full display each year on Bastille Day, July 14, when the French Armed Forces parade down the Champs-Élysées. Tens of thousands of enthusiastic spectators watch this fascist show of might. The loudest cheers are always for the stately column of legionnaires arrogantly bringing up the rear. The Legion was also as good at singing as they were at fighting, a boys' choir composed of tattooed men with husky voices. Singing and parading have always been the artistic expression that bonds men into brothers. One could hear the rich tone of bellowing legionnaires approaching miles away. As food in a restaurant resembles the skill of the chef, so too did our tenor and precision reflect the pedagogy of our handlers—by any means necessary.

A trumpet heralded our entrance to the courtyard. 28 chanting *engagé volontaires* marched in rows, following behind our *cadres* and lieutenants. Not a single soul flubbed the synchronized perfection.

"We the damned of the earth.
We are the wounded of all wars.
We cannot forget,
A misfortune, a shame, a woman we adored.

We have hot blood in our veins.
Madness in our head, and a heart of sorrow."

When we reached the middle of the courtyard, we halted and simultaneously turned to the left, just before the watchful eye of the 4e RE colonel who stood before the waving tricolor. With one loud final step, we all froze to attention. With one hand behind our backs, the other hung at our sides, holding our *képis*. Because we were arranged in order of height, I was given the honor of experiencing the spectacle up front and center.

The trumpet sounded again, and our captain marched in wearing the same camouflage gear as his men but with epaulettes with four gold stripes on his shoulders. He spoke with a crisp, loud voice that echoed off the castle walls. His words were tinged with spiritual notes: "*Engagé volontaires*, you have fought the good fight, finished the race and upheld the traditions of the Legion. You have earned the right to stand here today as legionnaires." Still standing to attention, we then reverently recited every word of the Legion's *code d'honneur*.

In attendance were no family members, no girlfriends, and no fanfare. We had nobody but ourselves, professionals, brothers by spilled blood—not birth. I didn't want anybody to mourn my death. Thanking us for our heroism was akin to thanking strangers for paying their taxes—it's merely what we do. The trumpeter blew two notes, and in prefect synchronization, we lifted our *képi* with our right hand and placed it squarely on our heads. This movement had taken hours to perfect, but we pulled it off flawlessly. We then saluted the colonel with the flick of our open palm to the forehead and back to our sides with a slap. He formally returned the gesture. The trumpet notes then danced around the courtyard, and we finally shared some informal bonding. I was now, a legionnaire.

With the Farm and a bitter winter behind us, we returned to Castlenaudary, which was slowly being transformed by the arrival of spring. The trees filled out in multiple shades of green and the days grew longer.

The second phase of *instruction* was more centered on the practical skills such as musketry, land navigation, and tactics. Though France hadn't defended against gas attacks since WWI, we trained thoroughly for it. As the days of fighting in trenches had ended, the Legion was now a mobile counter-guerilla force. Terrorist events had transformed our training, now with an emphasis on close combat.

The runs continued, longer and faster, but we were in excellent shape, and I actually enjoyed the morning workouts. We were now provided with more calories and gained back some muscle mass. However, the bullshit didn't cease but merely changed in nature. In between lessons, we were left to the entertainment of the corporals getting their sadistic jollies. They delighted in ordering us to paint every square inch of the corridor black with our own boot polish. We stood at attention while the corporals looked for any unpainted spots. Once they were satisfied with our work, we had to somehow remove any traces of black polish on the tiles, grout and all.

I preferred swift violence over mindless bullshit. It was quicker, and we could get on with more important things. With legionnaires used as their personal hand puppets, the corporals competed with each other as to who could come up with the most humiliating tasks. A legionnaire was once turned into a human cuckoo clock for a full 24 hours, forced to sit on a locker and shout out the time every fifteen minutes. Another had to stand at attention, also for 24 hours, in front of his bed. I saw him fall asleep standing up only to wake up when his head hit the bed frame as he collapsed. *Caporal* Bordon's creativity surpassed even that of the most skilled Spanish inquisitors.

Though I was living under my *nom de guerre*, there were no outstanding security restrictions on me. On Sunday afternoons, when we were occasionally given free time, I often spoke "off-the-record" with *Caporal* Basil. We kept our informal friendship under wraps, but like all things, breaking or bending the rules was *de riguer* in the Legion. *Caporal* Basil even passed me his cell phone to call my family, to announce that I was alive and now a formal member of the famed French Foreign Legion.

But then one morning I woke up in a cold sweat, feeling uneasy and unable to get back to sleep, even after being starved of rest. I stepped silently to the balcony and stared out onto the parade ground, looking past the barracks, taking in the deep quiet but feeling that something beyond the Legion was wrong. As my *section* queued for baguettes and coffee hours later, *Caporal* Basil approached somberly.

"You got a text from a friend back home. It sounded urgent." Benno, who'd motivated me to join the Legion, was now desperate to speak with me. This didn't bode well. *Caporal* Basil slipped me his phone, and I went to a secluded area. I left my breakfast uneaten.

"Hey man, tell me something positive. What's going on?" I hoped that my cheerful demeanor would bring good news. It wasn't to be.

"Alex," he started with a tremble in his voice. "Something happened..." My heart stopped.

"Just tell me, damn it," I said as I prepared for the very worst of eventualities. I heard Benno swallow.

"Jannie's dead. He was killed in a car accident. He wrapped himself around a pole."

"Oh God, no...please, not Jannie..."

"It was fast. I don't think he suffered."

My existence in the Foreign Legion suddenly felt meaningless. I was now back in Johannesburg.

"Alex, are you still there?"

Benno's voice echoed in my head from a distance. I began trembling from rage and grief.

"What the fuck am I doing in this army, playing cowboys and Indians? I need to get back home," I fretted.

"Mate, it'll do no good. Jannie's gone. He was proud of you for making it in the Legion. Death never comes at the right time. Don't throw it all away now. Do it for him."

"I need to be alone," I said and hung up.

I went through the rest of the next few days in a haze, merely going through the motions and trying my hardest to get through—or be thrown out. I put up with the pushups, punches, and hazing with little more than a comatose look on my face. The entire *section* knew that something was horribly wrong with me.

The nights were the worst. Jannie visited me in my sleep, chasing me on the rugby pitch, trying to buddy-up his big brother. "Don't forget about me," he'd say with his broad playful grin. I'd wake up several times in the night thinking he was standing next to my bed. Was it all a bad dream, or was he really gone? I remembered the pain of losing my father as a toddler, and not knowing why he was no longer at the breakfast table giving us our daily pep-talk before school. Jannie's loss revealed that I had never learned to let go after losing a loved one. I wanted to cry, but Jannie deserved more than my salty tears. Instead, I buried my sadness deep in my gut, to germinate and grow into an irrepressible fire.

But providence has a way of intervening just when all hope seems lost, for our *section* was just about to undergo our final tests. I channeled my anger into those tasks as if it might bring Jannie back. We were meticulously tested on every aspect of our training, from French to the FAMAS. One of the penultimate tests was how fast we could run eight kilometers in full kit and with a weighted rucksack. The commando run was perhaps the toughest to pass. It was a means of selecting those most suitable for the elite parachute regiment. The previous year, one poor devil suffered heat stroke during this run and ended up spending a year recovering.

When the time came to run, I had to prove myself invincible. I tightened my military belt and lugged my rucksack, filled with enough gear for a three-day mission, onto my back. I grabbed my FAMAS and green beret and headed to the maroon courtyard with my *camarades* to receive instructions.

Unconcerned about the near death of a recruit in the recent past, the *cadres* deliberately started us off at high-noon. My feet soon became red hot embers. It was torturous carrying that load, feeling it grate the skin off my back, but the pain was nothing compared to what my soul suffered after losing Jannie. Many gave up or didn't finish in the maximum allotted time. With cramping quadriceps and my spine painfully misaligned, I finished ahead of everybody else. My days plowing through men or wrestling them to the ground on the rugby pitch paid off.

In the weeks that followed, I attained one of the highest scores in marksmanship. For the Cooper Test, a 12-minute dash, I ran better than any burly Guinness-swilling rugby player in history, completing a decent 3100 m. I was ever grateful to my parents for motivating me to push through my studies since I got high marks for my near fluency in French. For our sandbag lift-and-sprint test, I imagined tossing a defender on my shoulders and rushing to the try line. I was able to operate the radio with ease and coordinate a *section* level counterattack. To me, leading men into battle was no different than motivating teenagers to break through a defensive line near the goal posts.

A man's best effort was always good enough for my father, and I never imagined I would out soldier the Spetznaz or my Yankee jarhead *camarade*. But to my amazement, I finished at the very top of my class! The Legion congratulated me with a low-key celebration, a certificate, and a handshake from our captain. I'd gotten used to working in a thankless job, but that simple gesture of gratitude from an army that had taken so much from me felt damn good. It pays to win, even though I would have traded it all to hug my brother Jannie goodbye before he left this world.

Finishing first meant that I was summoned to the *sergent-chef*'s office to discuss which regiment I preferred. The entire *section* queued in their order of placement. I passed the Harrak, who turned out to be a good bloke and slapped him on the back. I fist-bumped Daniels who finished in the top five. I then approached Marcos, the born natural, and gave him a warm hug. He was a better man than me in so many ways and was so close to abandoning his dreams. "You saved me back there," he said. "Thanks." I knocked on the *chef*'s door and crisply presented myself.

"You finished well," he started as he opened my file. "It is rare to see an Anglophone with such determination. As you know, you have the privilege of choosing where you would like to be posted. Your other *camarades* will be shipped off to where the Legion needs them."

"I want to go to Africa, sir, the Legion's original home."

Djibouti was home to the 13e *Demi-Brigade de la Légion Étrangère* (13e DBLE), created in 1940 as part of a constituted unit of the Free French Forces in WWII. It consisted of one infantry unit, one engineering unit, and one armored squadron. The 13e DBLE was involved in most French campaigns since then. Its strategic location in the Horn of Africa gave France a degree of control over the Red Sea and Suez Canal. As in many overseas units, legionnaires there could earn a small fortune.

The *sergent-chef*'s eyes narrowed, and he responded tersely.

"You cannot go to Djibouti."

"Sir, I was under the assumption—" I said quickly, but he interrupted me in mid-sentence.

"Because you are a flight risk. You are from Africa."

I was tempted to ask him if he'd ever looked at a map to see that there were several war-ravaged countries between the horn of Africa and my home. It would have been easier for me to get to Johannesburg from Antarctica than from Djibouti. But I didn't dare protest. The Legion simply had its way of doing things.

"Not that we don't trust you. We appreciate you Anglophones, but you do have a history of not sticking it out."

"Honestly, I recently thought about quitting. But now I have even more reason to honor my word. I would like to join the parachute regiment."

The *2ᵉ Regiment Étrangère de Parachutistes* (2ᵉ REP) had a reputation of being the cream of the crop. Based on the island of Corsica, desertion was believed to be impossible, which also gave the corporals free rein over new legionnaires. It was part of the *Force d'Action Rapide* (FAR), which meant legionnaires were on permanent standby to intervene anywhere in the world within 24 hours. The 2ᵉ REP was broken up into four companies, each with its own specialization in night, mountain, urban, or aquatic warfare. Even within this elite regiment was the "best of the best" unit, the GCP (*Groupement de Commandos Parachutistes*) pathfinders, open only to corporals and above.

"I'm afraid you cannot join the parachutists either," he fretted. "I know your dossier and your past as a rugby player. I even know the name of the doctor who operated on your bad knee. If we drop you behind enemy lines and you happen to reinjure yourself, we don't have the money or resources to rescue you. Now, what is your third choice?"

"Well, I didn't join the Legion to enjoy the beauty of southern France. Send me to the Amazon." At that stage, I didn't really care.

"Good choice!"

I now had a reason to smile, honored for being assigned to the 3ᵉ *Régiment Étrangèr d'Infanterie* (3ᵉ REI) in French Guiana "Guyane" straight out of *instruction*. Most legionnaires had to reenlist for that privilege. This dark horse cohort was steeped in rumor and mystery, notably with regards to the grueling Advanced Jungle Warfare Course. I expected Guyane to fill that gap left by my recent disillusionment during *instruction*. I joined the Legion to expose my body and spirit to the most extreme conditions. The Amazon would finally fulfill that expectation.

Escaping the Amazon

Still in Castel, before I was slotted to ship out to the northeastern coastline of South America, the entire Legion celebrated its most sacred holiday, Camerone Day. In the 1860s, France was attempting to extend her empire into Mexico and sent in the Legion. Many deserters at that time ended up fighting in the American Civil War. Towards the end of April 1863, French troops stationed in Puebla Mexico were running low on supplies. The Legion was selected to guard a special relief convoy. Only sixty-two legionnaires, led by a one-handed *Captaine* Jean Danjou, were available to participate in the mission. There were doubts that such a small company would be able to police the party, but *Captaine* Danjou knew that his legionnaires would fight to the death to accomplish the task. He didn't know how true this would be. Along the way, near the town of Camerone, they were surrounded and attacked by two thousand Mexicans, and a daylong battle ensued.

Five times, the superior Mexican forces called upon *Captaine* Danjou to surrender, and all five times the answer was: *"Merde!"* Echoing that spirit, in 1944, when American commander Anthony McAuliffe was made the same offer by the Germans, his response was: "Nuts!"

The captain was shot dead during the battle, and only three legionnaires survived. While the fighting raged, the convoy slipped past, saving the troops in Puebla. After fighting with such extraordinary bravery, the lives of the survivors were spared by the Mexicans. *Captaine* Danjou's wooden hand was later recovered and sent back to France as a relic of the Legion's heroism and sacrifice.

Every year, thousands of veterans in various cities celebrate. The main gala occurred in Aubagne, but each regiment and outpost participated in their own way. In the Legion, even those who die in a lost battle merit Valhalla. I didn't understand the spiritual ramifications of this death cult at that time; I just knew that this was the first occasion in months that we were allowed to drink.

Camerone Day was one of the few times that barriers of rank were broken, with sergeants, corporals, and legionnaires mingling with each other freely. Yet we were still young legionnaires in training.

One thing that legionnaires were good at was turning any celebration into a massive party. Marcos was showing off some smooth capoeira moves. Daniels was already drunk off of a fifth of the most excellent Kentucky bourbon, which cost him several weeks' pay. Trying to prove ourselves amongst a multitude of nationalities, with alcohol thrown into the mix, there was bound to be trouble.

Just after I passed out in bed from drink and exhaustion, I was woken by a commotion outside in the square. I put on my military trousers and a T-shirt, splashed water on my face and went outside to see what was going on. A dark cloud permeated what had been a joyous atmosphere. I heard loud voices and sirens. The gendarmes and civilian police were present beyond the regiment gates. Near the company entrance was rivulets of blood, presumably from a drunken punch-up.

The authorities were questioning men who were still in their underpants and those who could barely stand. Other legionnaires looked out from their balconies with stunned looks on their faces. Yet, nobody quite knew what had occurred. And then I spotted Marcos, who was not a heavy drinker, on the far side of the square with Daniels.

"What the hell?" I asked.

"Something terrible happened last night."

"Not the typical ass-beating but something really bad?"

"You know that creepy central Asian fuck who has barely spoken more than three words during *instruction*?"

"Yeah, I steer clear of him."

"He sodomized Myshkin in the toilets with a broomstick. There's blood everywhere."

"What?"

"We'll lynch that Genghis Khan piece of shit," Daniels said with rage in his eyes.

"The pussy fled, and they can't find him," Marcos continued. "He singled out the weakest teenager in the *section* and had his way with him. Poor kid nearly bled out, and now his insides are torn the fuck up. We'll probably never see either of them again."

I wondered who was to blame. After the constant harassment and abuse, now mixed with alcohol, the culprit snapped. Did the Legion turn Neanderthals into gentlemen or vice versa, or did it simply allow the sick men of the world to channel their perversions? Was it *Caporal* Bordon's incessant hazing, the Legion's fanatical discipline, or deeply repressed feelings of inferiority and helplessness that manifested itself in cannibalism? Why was my brother Jannie dead, and this sick rapist still alive? I couldn't answer that question.

And who among us was without sin? I struggled to believe that all wrongs were equally repugnant before the Almighty. God may have forgiven him, but I did not.

The following day, it was back to sweeping, polishing, and scrubbing, as if nothing had happened. Marcos was to ship to 1er *Régiment Étrangèr de Genie* (1er REG) in Laudun and Daniels was first in line to go on a mission to Afghanistan with the 2e REP. "As long as I can keep killing those hajji motherfuckers, I'm straight. Oorah!"

EDEN

A mission once given to you becomes sacred to you, you will accomplish it to the end and at all costs.

Code d'honneur

To ship us off to Guyane, I imagined that we would be packed into the back of a four-engine Transall C-160, the fuselage ruggedly bare, with steel benches running to the cockpit, red netting to hold onto through turbulence. Everyone would be sitting in silence amidst the noisy turboprops, faces covered in camouflage paint. Instead, I found myself in the very modern Marseilles Airport waiting to board a sleek Boeing 747-300.

"Budget cuts," I said to Marcos, "the Legion is the first to be sent and the last to be funded."

Dressed in civilian clothing, with French passports in hand, we checked into our flight to South America. "Hey, if you're going to desert, now's the time," Marcos added in earnest. We discovered later that someone *did* exactly that. A soft spoken Bolivian herdsman ducked out to take a piss and never returned. With a wad of Legion earnings, he paid cash for an immediate flight home before anybody took a headcount. We admired him for having the brains and balls to pull that off.

It was invigorating to walk through an international airport as a legionnaire and not a civilian. With our shaved heads, tight jeans, and ratty tank tops, we could have been extras in a Frankie Goes to Hollywood video. Civilians knew who we were and steered clear of us, either out of respect or fear. A boy ran up to me and asked if we were being sent to Afghanistan, Algeria, or Chad.

"I'm going to the Amazon to protect your rockets."

He returned a perplexed look. I posed with him for a photo, thanked him for his good wishes, and walked through passport control to board. After months of sleeping in the snow and mud, I was uncomfortable being cozy in my plush, warm seat. We managed to traverse the Atlantic with no civilians beaten or flight attendants sexually assaulted.

My first and most lasting impression of Guyane was the immense, suffocating heat and humidity, which hit me the moment I stepped onto the tarmac. As we made our way by bus to Kourou and crossed the Cayenne River Mouth, I realized that I was no longer in Kansas. I gazed across at the estuary, which seemed to be swallowing the mighty Atlantic. The flat, lazy beach town of Kourou felt deceptively like a holiday resort. The abundance of palm trees and people walking dozily in the streets was a pleasant welcome.

Guyane is a small country nestled in the northeastern corner of South America. Native Americans inhabited it for centuries until Columbus brought the influx of European influence and hunger for land. Subsequently, France tried several times to seize ownership of the region but failed. A hundred years after finally gaining control of the territory, France's King Louis XV sent thousands of settlers to build a colony there. The attempt was horribly unsuccessful, with thousands succumbing to tropical diseases and local hostility. Those who did survive fled to a trio of tiny islands offshore.

90

As a failed colony, Guyane became a naughty corner for revolutionaries, unwanted politicians, journalists, or slaves that France needed to dispose of. In the mid-nineteenth century, France passed a law that anyone with more than three convictions for theft would be sentenced to Guyane for six months.

Well into the twentieth century, France ran the most infamous penal colony in history, Devil's Island. Opened in 1852, it was notorious for its harsh treatment towards criminals, who were deported there from all parts of the French empire. Laws required convicts to stay in Guyane after completion of the sentence for an equal span of time. If the original sentence exceeded eight years, they were forced to stay for life and were merely provided an unworkable tract of land on which to settle. A limited number of women, commonly convicted of infanticide, were also sent to Guyane, with the intent that they marry freed male inmates to aid in the development of the colony. Of course, the islands also held the most hardened criminals. Prisoner-on-prisoner violence was familiar and tropical diseases rife. Sanitary systems were scarce, and the region was mosquito-infested. The only escape from the island prison was by water, and few convicts succeeded in doing so. The vast majority of the eighty thousand prisoners sent to Devil's Island never made it back to France. Broken survivors who did retold of the horrors. They often scared other potential criminals straight. This system was gradually phased out and completely shut down in 1953.

The Dreyfus Affair was a political scandal that divided France from 1894 until its resolution in 1906. Often seen as symbolic of modern anti-Semitism, it remains as a striking example of a complicated miscarriage of justice, where a crucial role was played by the press, intellectuals, and public opinion. The scandal began in 1894, with the treason conviction of Captain Alfred Dreyfus, a young Jewish artillery officer. Given a life sentence for allegedly communicating military secrets to the Germans, Dreyfus was imprisoned on Devil's Island. Apart from the guards, he was the only inhabitant on his island and was held in a tiny stone hut. Worried by the risk of escape, the commandant of the prison was exceedingly harsh. Dreyfus became sick and shaken by fevers that gradually worsened. An artists' revolt in Paris and the publication of Zola's *J'accuse…!* finally led to Dreyfus' freedom and exoneration.

Henri "Papillon" Charrière was convicted for murder by the French courts. He was known as the author of *Papillon*, a memoir of his horrifying incarceration in, and death-defying escape from Guyane. After Charrière's final release in 1945, he settled in Venezuela where he married a local woman. He was subsequently treated as a minor celebrity, even being invited frequently to appear on local television programs. He finally returned to Europe, visiting Paris in conjunction with the publication of his memoir. The book sold over 1.5 million copies in France alone and was made into a major film starring Steve McQueen.

In 1964, France finally found an economic use for its colony and set up the Guiana Space Center, which brought economic development to a population scarred by centuries of greed and bad planning. The 3^e REI took garrison at Kourou in 1973. In the Legion's pioneer tradition, the regiment immediately marked the territory and blazed a route towards the east, which now links Cayenne to the Brazilian frontier. The primary mission of the regiment was to secure the Space Centre, specifically the exterior zones for each launch. It also protected the sensitive surrounding installations from low-altitude aerial threats. Every Legion mission required the deployment of three combat companies.

The subsequently formed Jungle Training Center in Régina is the world's premier survival and tropical warfare school. The Legion's grueling two-week Advanced Jungle Warfare Course is known to break even US SEALS and British SAS contingents. Combat *sections* conduct intervention operations lasting up to several excruciating weeks. The Legion boasts of the ability to go deeper into the jungle and survive longer than the local indigenous peoples.

The 3^e REI is projected in the Caribbean South America zone as a prepositioned operational force capable of intervening instantly, as in 2004 during Operation Carbet in Haiti. Far away from the watchful eye of any human rights groups, in the 2000s, the Legion pushed to thwart illegal mining and drug trafficking activities deep in the jungle. This objective became permanent and was later reinforced. Such was my lot as a ranker legionnaire. I joined the Legion to save lives and experience adventure, not to pistol-whip smugglers and ensure satellite television connections for Parisian civilians.

We were indeed supplanted in God's Garden of Eden, yet I wasn't sure if France's legionnaires and rockets were the uncorrupted Adam or the serpent. Westerners are prone to glamorize man in his natural state. There is a fine line between paradise and madness, and men free from the constraints of civilization became noble savages or simply savages.

I wore my uniform with pride and took on any assignment with relish and excitement—until given my first mission. A detachment was tasked with guarding the Space Center's perimeter before a launch. We bivouacked in a small section of grass between the outside fence and the river, and continuously patrolled in shifts. Boredom set in after a week of launch delays. We killed time by joking, sleeping, and drinking ourselves into a stupor. With no ammunition or real combat skills, we were a threat only to the mosquitoes.

Despite the monotony, terror was never far away. While we monitored our section, a dozen men dangled their bare feet in the slimy riverbank, while the others played cards on the grass. Still, others took an undeserved nap. But we let our guard down. I was trying unsuccessfully to sew a button back onto my uniform when I heard a gargantuan splash behind me. I swung around to see a kaleidoscope of flesh, teeth, tails, and arms tossing water in all directions. One of our men was in the water! It all transpired in a split second and a short distance from the shore. We were all petrified by the horrifying spectacle. Not a single soul had the courage or gumption to intervene in the battle between man and beast. Before I could shout, "Get him out of there!" blood stained the river, and all went silent. As soon as we snapped back to reality, our fellow legionnaire emerged from the water dragging an adult caiman onto the sand.

"Anyone hungry?" he asked, smiling from ear to ear.

I thought I'd met the most amazing characters on the Farm. I was wrong. Every *section* always had one guy who went beyond the pale. We all ate reptile for lunch that day.

The Legion has a patronizing maxim that legionnaires are happiest when they are working. This smacked of the notorious Diefenbach motto of *Arbeit Macht Frei*. When we weren't mopping perfectly clean floors or watering petunias in the colonel's personal garden, we battled boredom. War movies are based upon actions on the battlefield, not the drudgery of garrison life. For every hour a fighter pilot spends in his $35M jet, a hundred man-hours of maintenance, refurbishing, replacing, and testing the aircraft were required. The days, weeks, and months wore on, dotted with sojourns into the jungle. When not on missions, I dipped into a state of depression.

During the Legion's conquest of the Sahara, thirst, heat, and the NCOs' brutality often led to violence or suicide. Legionnaires were known to bayonet a *camarade* for no reason, or dash out into the desert naked to be captured and tortured to death. This form of psychosis became known as having *cafard* (cockroach). Legionnaires are usually not even aware that they are suffering from it. When an old hand grumpily says, "*J'ai le cafard, plein les couilles*," he's telling all others to stay away from him for their own safety. *Cafard* fed into the mob mentality, was contagious, and lead to mutiny or mass desertion. Many men who succumbed to it were model legionnaires who rarely even muttered a word. The rapist in *instruction* couldn't have been in his right mind when he committed such a horrific act. The only cure for *cafard* was whores, drink, and that one thing every depressed legionnaire prayed for—combat.

Our free time at the REI "ray" was usually spent getting drunk in the foyer. If one had the will and energy to press his uniform, he could do the same in Kourou with the locals. But alcohol was pricey in town, and an entire month's pay was often blown on one night of bacchanalia. Even though I was one of the more clean-cut types, I was beginning to enjoy booze more than I should have. Buddhists, yogis, and Christian monks spent years mastering their minds to distort or enhance perceived reality. For us, doing so was as simple as cracking open a case or ten of Kronenbourg. Alcohol either made me love the Legion or curse the day I was born. Some legionnaires graduated to harder drugs. Commanders turned a blind eye to such activities, seeing it as medicine to stave off a revolt.

Like university students on spring holiday, it was easy to engage in uncharacteristic behaviors when one was in a new and anonymous environment. Yet I was amongst men who hadn't seen a woman for upwards of six months. Away from the strictures of family, custom, or European law, it was all too simple to surrender to the sirens' temptation of prostitution. Unlike some of my colleagues, I had a healthy attitude towards women. I was raised by strong sisters and an even stronger mother. I wasn't too tall but had a decent face not scarred by weekly punch-ups with the police, so I never had to try too hard to find female affection. In any case, I believed that any intimacy had to be earned the old fashioned way.

"Come on Darklay. It's what soldiers have done since war was invented," I was egged on. "Are you saying that you prefer chatting up some illiterate woman, buying her a gift, and inviting her for a candlelit dinner in the hopes that you might be 'rewarded'? Just get it over with, mate."

"Hey, knuckle draggers, that whore you shag is somebody's daughter or mother. If you really want to help her out, buy her kids some shoes."

I felt uneasy whenever the magical hour came when the boys decided that it was time to get laid. I'd politely slip away, drink alone, or catch a taxi to the REI. On my way back one evening, just past our favorite watering hole, Las Iguanas, I saw something bobbing about behind a bush. As we drove past, I spotted a corporal from a neighboring company with his *képi* on backwards, a woman bent over in front of him, and a Kronenbourg in his right hand. He took a sip and slapped the prostitute on her ass. Her face was emotionless, as if she had lost her soul decades ago, and was now just a wandering piece of flesh to be exploited by legionnaires. My heart broke for her. In light of the locals' permissive attitude towards prostitution, I wondered if these women, in some masochistic way, preferred that life. Until very recently, the Legion ran its own brothels. They were finally disbanded, not on feminist or human rights grounds but because the wife of REI's commander was a practicing Catholic and would have none of it. A 1980's documentary narrated by Algerian-era legionnaire Simon Murray included a touching conversation with a retired prostitute who stated that she owed her life to her legionnaires, and longed to be buried next to them when she passed.

It was easy to judge, yet I saw prostitution as an act of desperation by both parties. Sex with a prostitute was merely a proxy for power and status for men who had neither.

Historically, one district of any overseas garrison town was strictly off limits to legionnaires, under a penalty of a month's imprisonment. The mythical *village nègre* was home to every sort of disease and vice. Prostitution in Guyana had a dark history, tinged with notes of conquest and slavery. Discerning between temptation and intellectual curiosity was difficult. I walked through this quarter once and was taken aback at the conditions, appalling even by local standards. The main street was a narrow little alley. I could almost touch both of the walls on either side. The low houses were essentially ruins, and rough holes substituted for windows and doors.

Songs, cries, and shrieks filled the air. This was the last stop, even for women from more prosperous lands. In the corner of a hovel, I faintly saw a thin and sickly blond woman, who may, in fact, have been French. Legionnaires of yesteryear went on missions through the *village nègre* to rescue white slaves, European women who were kidnapped by Arab traffickers from French or Spanish beaches and spirited away to North Africa. By the time those women were found, they often resisted being "rescued." Here, a Brazilian, in whose face her unpleasant life had cut deep lines, sat in a torn silk dress on the bare ground. She was too exhausted or high to speak, and merely invited pedestrians to come inside with a wave. Near her, a local black woman with a robust figure lay stretched out almost naked. Young Colombian girls crouched next to them. In the midst of this misery moved the lowest social classes of Guyane. Men who labored mercilessly under the sun during the day spent their pay with equally destitute women.

Australian Dave Mason who served in the mid-nineties wrote about an incident he witnessed in Djibouti. He walked in on another legionnaire having sex with a local prostitute. Mason was offered to "have a go" with the girl but declined. "It's true, you know, if you punch them here," said the other legionnaire, whereupon he gave the girl a quick, hard punch to her side just below her ribcage, "their guts'll spasm around your dick."

It wasn't too long after I arrived that I met the great Connor. Old legionnaires either finished their contracts, committed suicide, or were sent to the knackery in France—the retirement home in Puyloubier, also known as the former Nazi B&B. As such, the REI received a fresh shipment of cannon fodder every two weeks, via the same flight on which I had arrived. One evening, I was back at Las Iguanas with my faithful bunch of drunks and perverts. At the usual time when they ventured off to "donate money to impoverished women," and as I was closing my tab, a group of newly minted legionnaires walked in through the door. In the center of the pack stood a fit, young man who towered a head above the rest—Melville's Handsome Sailor. He had to duck to avoid the sputtering ceiling fan. I expected him to don a white ten gallon hat, and announce to all that he "was here to clean up this town." The others sat at the bar beside him, on both sides, and seemed honored to be in his presence. His entourage was well behaved and immune to the company of the usual prostitutes. Curious, rather than call it an early night, I decided to buddy up to them just to see what all the bonhomie was about.

"Darklay's my name," I said as I extended my hand.

"Connor," the lofty lad responded as he rose to his feet, and returned a firm, confident shake. "We're all mates. Join us."

I put down my Kronenbourg and asked what one is never supposed to ask. But it was too irresistible not to: "So why'd you enlist in the Legion?"

"Well," Connor said with a big toothy smile, "I had few options but wanted to do something big with my life." His intentionally vague response only made me more curious.

"I have no idea what you mean by that," I responded with a tinge of cynicism.

"Australia's a large open country, you know. It's easy to get bored," he told me with a twangy accent. "There's nothing more to it, mate."

I didn't press him further. Such details ooze out only when the time is right. After a few more Kronenbourgs and some earned trust, I nearly connected all the dots. He had left a promising swimming career in Australia to join this gang of roughs. His dream was to one-day win a medal for his country. Rumor had it that he was injured right before the Beijing Olympics and couldn't compete, which ultimately led him to the same place as me.

The Legion always posted its top swimmers to the Amazon regiment. Connor was just another of the fascinating, mysterious, and often sad souls who ended up here. We very soon became close *camarades*, not only because we spoke the same language, but because Connor was the amiable, all-round Aussie golden boy of the regiment. Even the NCOs took a liking to him. I was taken aback the first day he and I were put on landscaping duty. As we painstakingly walked around, bending over pulling weeds, everyone who passed Connor greeted him with a warm handshake. One particularly unpleasant sergeant gently approached him and sought his counsel about a woman in town. I couldn't have cared less about such trivialities, but Connor was always eager to listen, to not pass judgment. Naturally, I looked up to him as a mentor, not only because he felt like an older brother, but also because he was a hulking 6'11" man. He towered above my barely 5'10" frame, and most others in the REI. He had a smile as broad as he was tall, and the patience of a saint, albeit an angel who could push back the drinks. We spent many hours bonding over Kronenbourg, and dreamt about the elusive adventure that lay in the Amazon, and even of our plans to start a business together if we ever made it out alive.

The rainforest was a suffocating place when one used up days wandering its endless paths. On my first two-week mission along the Brazilian border, I realized how insignificant even the mighty Foreign Legion was in comparison to the multitude of life teeming under the calm tree canopies. The ground was permanently wet and covered with decomposing flora and roots. It was eternal night, for the sunlight was absorbed by the canopy above us. The Amazon houses a third of the world's birdlife. Mammals, reptiles, and insects were also in abundance. We walked past massive moss-layered tree trunks, oblivious to the fact that fifty meters overhead, their leaves were producing twenty percent of Earth's oxygen. On the floor of this ocean of life, in our denim combats, the stifling heat was unbearable.

With a growing population and rising standards of living in the region, mining, logging, and cattle ranching endangered the Amazon. But the Legion wasn't stationed there to further philanthropic, environmental causes. We were there to curb illegal trafficking of drugs, gold, and other contraband from Brazil.

On one particular mission, we'd been blazing a path for most of the day when a bare-chested man darted from the undergrowth directly across our way—the naked boy in Gethsemane—running himself into the hands of our *section* leaders. Our sergeant felt threatened and immediately rushed the suspect, punched him in the stomach, pulled his neck down in a Thai-clinch, and kneed him viciously in his face. The sound of his nose breaking was sickening. Blood poured from his face like wine. I was shocked by the brutality and had to restrain myself from intervening to protect him. We were then ordered to strip search the trespasser. He had small patches sown on the inside of his pant legs. I tore them off, and five irregular lumps of gold dropped to the ground.

Every legionnaire underwent the same training. "Depending upon how he is led, a legionnaire can be the best of soldiers or the worst of brutes," a twentieth century commander was quoted as saying. It was human nature to choose the path of least resistance. In examining wartime atrocities, such as the My Lai Massacre in Vietnam, psychologists estimate that 70% of a group would gleefully join in the carnage; 20% would simply look the other way, and only 10% would actively resist. I wondered if I were male enough to be in that latter percent.

It was sickening to see this grown man whimper and cry like a child. He pleaded in a mixture of English, Portuguese, and an indigenous language, but I roughly understood him: he had a family and was begging for mercy. He was unarmed, no larger than a teenage boy, and looked like he hadn't had a meal in days. I wondered whether his children would eat again if he didn't deliver this gold to his handlers.

His hands were tied behind his back, and he was interrogated for two days by our superiors, but it was far from over. We were ordered to cut up his shoes and burn everything he had on him. We obeyed without question. "But my feet…" he pleaded in tears. Even in military issue boots, our soles were shredded after a few hours of walking in the jungle. We aimed to get him to lead us to his accomplices. "These criminals never work alone," a colleague explained to me. The smuggler's face was now streaked with the mixture of dried and fresh blood and saliva. And then he finally broke and agreed to direct us to the others.

Completely naked but for the rope that held his bleeding wrists together, he guided us up to the crest of a steep hill. Our lieutenant then turned to me. "You stay here with him, and we'll go down. If he moves, shoot him, and then drag his body into the jungle for the animals to eat." Our hostage must have understood some French because he immediately lay down on the ground with his head in the mud and didn't twitch for three whole hours.

I joined the Legion to save lives, not to take them. I wanted to learn the noble warrior trade and to live by an ethical code. I had sugarplum dreams of rescuing nuns caught in the crossfire of a civil war, enforcing peace in the midst of tribal conflict or simply bringing food to famished children. But now I sat in the jungle, sweat dripping off my face, with my FAMAS in the back of a defenseless man's head. We were there to curb the illegal gold trade, but this man's helpless situation didn't conform to the amoral mandates penned by politicians. The directive to curb illicit mining was nothing more than a power play between regional powers, and as a result, sick entertainment for legionnaires on the ground. I was a coward for not helping this man, and for the first time, wondered whether I was cut out for this line of work.

I never knew what happened to the confiscated gold. When my *section* returned from its fruitless search, our hostage received another beating, "just to teach him a lesson," before they reluctantly let him go, naked and shoeless, into the brutal Amazon, his eventual fate known but to God.

Escaping the Amazon

On our march back to the REI, while I was still in a daze of confusion and rage, we came across a small herd of cattle, a strange thing to find in a wild and savage rainforest. A corporal casually handed us live rounds and ordered us to shoot them all. No explanation was given at the time, and nobody asked. Amongst us was a group of legionnaires who had just returned from Afghanistan; bloodthirsty and trigger-happy, they gleefully volunteered to slaughter the animals. Within seconds, mayhem broke out. Shots were fired at random, one after the other. The animals scattered, confused, trying to escape. They ran into tree stumps and stumbled over rocks and roots, splitting hooves and breaking bones. Rounds hit the cattle in the legs and sides—none of the wounds fatal. The sick executioners were torturing the animals and delighting in their agony. Some cows were shot twenty times before they died. Blood spattered against the green foliage, and it made the nearby creek flow red, foretelling the apocalypse. The cows moaned in a uniform aching tone. When the killing was over, the lieutenant counted the rounds that had been shot—surely for a signed and stamped inventory report accompanied by a spreadsheet to be delivered to the colonel—two hundred rounds to kill eight animals.

    I spent many days on a farm as a child. I also enjoyed a nice steak on occasion. But Pa taught us never to order meat unless we were going to eat all of it. Even grizzle and bones were fed to our dogs rather than be tossed into the bin. Animals died so that humans could live, and I revered them for that. God may have given Adam and Eve dominion over beasts great and small, but they were not put on earth for our sick pleasures. Privilege came with responsibility, and now we had destroyed a years' livelihood for several families, all because those farmers were forbidden to graze their cattle on that patch of land.

In my travels through Southern Africa as a youth, I remember a conversation I had with an indigenous San bushman. "I respect the game that I hunt because they are very clever. From an early age, the animals recognize man—the hairless ape that uses steel tools. They know that when they spot us, the encounter will not end well." In *Devil's Guard* by George R. Elford, recounting his exploits as a former Nazi SS turned legionnaire in Indochina, the author writes about a conversation with a big game hunter. The protagonist quips: "I hunt the most intelligent but cruelest animal of them all—man."

In a subsequent mission, my *section* found itself in the protected area of Saul, in the heart of Guyane. It was impenetrable even by Amazonian standards, and entirely inaccessible by any land vehicle. We had to be flown in by military plane. We were to establish a base camp there and then branch off on smaller missions on a bi-weekly basis.

The Legion had expelled hundreds of other illegal miners from the area the year before, and our mission was to ensure that the mines remained abandoned. Although there was a river close by, we had to painstakingly hack our way through the vegetation, foot by foot, simply to get to the site we were policing. We spent five exhausting days doing precisely that. When we arrived, we confirmed that the mines were untouched, evidenced by the overgrowth, and there was no sign that anyone had even approached the site. With that mission accomplished, we turned back, and I finally lost my sense of humor.

"What the fuck was the point of this?" I snapped at our lieutenant, throwing my rucksack on the ground. So much as eyeballing an officer in the Legion was unheard of, and in that location, could have easily resulted in my being buried in a shallow, muddy grave. But after busting my ass on countless stupid missions, I didn't care. "We risked our lives walking for a week getting here. We could have made it by boat in one day!"

Our lieutenant was undoubtedly descendent of the WWI generals who casually sent waves of men to be slaughtered, noted that their strategy didn't work, and then launched yet another wave. He stared at me, and chose not to respond to my gross insubordination—this infuriated me even more. Instead, he leaned back onto a tree stump, lit a Gauloises, and smiled smugly. "The boat is loud. If they heard us coming, they'd flee."

Enraged, I stood up and grabbed my machete. "Darklay, what the fuck!" my *section* gasped. But I simply banged it against the closest tree. The sound ricocheted throughout the forest. "We've been making this racket for days!" I threw down the machete and spat on the ground. I made my point, but I paid a steep price for it. I was tasked with using that same machete to clear the path for the entire detachment behind me as we forged yet another virgin trail through the Amazon.

On day two, we stumbled across an anaconda. The legionnaires gathered around, staring at it as though it were an alien. It was young and must have just eaten, as it was unusually docile. The creature was as long as a man is tall and its coloring still vibrant. I was fed up, tired, and sure as hell not going to stand idly marveling at a fucking reptile. I reached down to grip its mouth, taped it shut, and shoved it into my rucksack.

Without a word, I kept walking, too stubborn to admit that I now had to lug an extra twenty kilograms through the bush. Two days later, my *section* prepared a special dinner, and my load was twenty kilograms lighter. We sat around a fire and one pot, to which we added copious amounts of rum, which the closet alcoholics carried in their canteens instead of water. The Asians in our midst cooked a batch of sticky rice and served our meal on banana leaves. We ate with our hands and enjoyed knowing each other on a deeper level. Like a fine Bordeaux, the snake tasted of the surrounding jungle *terroir*, fishy in texture with a luxurious finish. The vertebra, which makes up most of the reptile, made it difficult to eat, as we had to pick the meat out from between the bones. Despite the stupidity and drudgery, moments like that mitigated my contempt for the Legion. It was what it was.

Our commander knew our minds better than we did, and with the whiff of mutiny in the air, we were given a new task. Before embarking on our second mission to Saul, I felt a particular disquietude. We marched with a relatively large *section*, many of whom were newly arrived *jeunes*. The atmosphere was thick as we descended a steep slope that led to the river below. The foliage was so dense we couldn't see more than a few meters ahead, but it thinned out as we approached the bank. And then we suddenly heard voices in Portuguese. We all immediately assumed the same thing—Brazilian smugglers.

Our sergeant stopped dead in his tracks, turned around, and hand-signaled us to take cover. We crouched down in the bushes awaiting further instructions. Without muttering a word, he divided us into two groups and communicated that we were to lock-in the infiltrators from either side, down in the riverbed.

As we slowly approached the voices, I was careful not to slip on the muddy ground and give away our position. I risked dropping my FAMAS since I had only one sweaty hand with which to grasp it. My free arm was the appendage that kept me from falling tits-over-toe into the mire and clay. Suddenly, one of the *jeunes*, unable to contain himself any longer, ran towards the suspects screaming like a banshee. He wanted desperately to put to use what he had learned from Rambo films and wasn't going to miss his opportunity.

The Brazilians took flight upon hearing that guttural howl from an unknown Amazonian beast. We now had no other option but to engage in hot pursuit. We were loaded with gear, and I clomped along as fast as I could when I reached level ground. Like the Gingerbread man, the three Brazilians flew over fallen trees with animal-like agility and split in different directions.

The corporal leading my group chose to trail the smuggler who headed upstream along the river. I followed, jumping over roots and ducking under low-hanging branches. The Brazilian changed course and sailed over a broad stream. The corporal, without thinking, jumped across the same brook, but his pack was too heavy, and he underestimated the span. In mid-air, he dropped like a rock, face first into the muck. I had only one thing on my mind. *Don't let them get away.* Without stopping, I charged across the stream, using the corporal as a stepping-stone—he never forgave me for that one.

My chest burned, but adrenalin kept me going until we reached a second, even wider creek—now I understood why the Legion tests for the rucksack run. Again, the Brazilian cleared it like an Olympic long jumper. I followed diligently, with the "Of course. You're my boy" spirit. When I approached the stream, I thought back to the time when I had easily leaped out of the sandpit in school. But hubris got the best of me, and it was my turn to fall face first into the mud on the water's edge. I was stuck, and the chase was over.

My struggle to get out of the mire only sucked me in deeper. It took five minutes and all my strength to finally pull myself free. I had new, hard-earned respect for quicksand. Caked in grime, I made a cursory search of the area, though I expected the Brazilian to be long gone by then. I walked up stream for a few meters in the same direction I'd been running. When I came around a bush, I suddenly found a shotgun pointed at my nose. It was so close that I could smell its distinct anti-corrosion oil. The bloke holding the weapon had a trembling finger on the trigger. *Fuck!*

In a reflex response, I immediately pulled up my FAMAS and held it to his forehead. I shouted in French, trying to intimidate him. I was bluffing—there were no rounds in my rifle. The rules of engagement in Guyane forbade carrying loaded weapons in a peace zone. Live rounds were available, but we could shoot only if shot at. I initially agreed with this rule's logic, but staring down the barrel of a loaded shotgun with nothing but an expensive fiberglass and steel club in my hands quickly changed my opinion.

We stood there, guns to faces, screaming in languages that neither understood. He could have easily killed me, and I knew it. Like a deadly Punch and Judy show, we both shouted at each other to drop his weapon. Eventually, my corporal—who had temporarily forgotten about my earlier disrespectful act—and another legionnaire heard my yelling and showed up next to me with their rifles pointing at the young Brazilian's head. Now he had three unloaded weapons directed at him.

A few more tense moments followed. His eyes pierced mine; his finger grazed the trigger. After weighing up the options, he finally dropped his weapon, thanks be to God, as I didn't want anybody to get hurt, or for me to be flown back to France without a head.

I feigned bravado among my peers, boasting that I knew exactly how things would play out and that it was all in a day's work. The remaining members of the *section* somehow caught the two other accomplices and bound them together. Even legionnaires not involved in the chase were gloating. But it was not within me or our *code d'honneur* to humiliate the enemy when he's down. Our sergeant tried questioning the smugglers but couldn't understand a word they said. Nobody in my *section* spoke Portuguese.

The principle of probable cause was best left to stuffy law school lecture halls. It never occurred to anybody, before we charged with weapons blazing, that the suspects might have merely been local fishermen minding their own business. Lavrentiy Beria, the Soviet Secret Police Chief, coined the applicable term "You bring me the man, and I'll find you the crime." We ignored any search and seizure protocol, raided their bags, and found nothing of significance. Yet we weren't about to let prickly things like facts interfere with satisfying our adrenalin-filled sadism. We then ordered the detainees to strip down to their underpants.

We threw all their belongings, clothes, shoes, food, and wallets, into a heap between two trees and set it alight. We stood and watched the three young Brazilians sob as they saw everything they owned go up in flames. There was no tribunal and no legal recourse for those wronged men. But the wrath of Paris would fall on any local who so much as looked at a French tourist the wrong way.

When everything was burnt to ashes, we simply walked away. I hesitated to look back. The sight of their slumped, naked bodies made me weep. Again, I didn't stand up for them, nor did I protest. For those poor souls, it was a seven-day walk to the nearest outpost, if they were lucky. The jungle was deadly even for trained troops with equipment, clean water, and tinned provisions. I never knew if they made it out of the Amazon alive.

It wasn't long before I saw the sad face of *cafard*. A company from the *2ᵉ Regiment Étrangère d'Infanterie* (2ᵉ REI) headquartered in the French postcard town of Nimes joined us for a four-month *secondment*. It was standard for our regiment to have on standby two permanent combat companies and one rotating contingent from the mainland. Being a legionnaire with no rank, the purpose of our missions was never communicated to us. We just went where the Legion needed us. My company was tasked with relieving the 2ᵉ REI, who was manning a remote post deep in the rainforest. We planned to bivouac alongside them before they departed the following morning back to Kourou.

When we arrived at camp, the morale was noticeably lower than expected, considering that they were about to return to *la belle Europe*. Being snatched abruptly from Provence and dumped into the jungle for four months, we clearly saw that the experience had taken an enormous toll on them. The men sulked and talked little. They merely sat around, shaved in the river, washed their clothes, or rolled Drum brand cigarettes. There was a breakdown in leadership, and nobody seemed to be in charge. We felt like the supply party arriving years late at the lost English colony of Roanoke. We sensed that something tragic had happened but couldn't figure out what.

As dark approached, the 2^e REI organized the usual sentry schedule for the night. *Jeune* legionnaires rotated on two-hour shifts and patrolled the surrounding area. The evening was unbearably humid, as usual, but uncharacteristically cool. The hair on the back of my neck stood up. A new sentinel was sent out at midnight. My section was wrapping up their conversations and settling into their hammocks for the night. But for some reason, my soul rested heavily in my chest, and I couldn't get to sleep.

Only moments later, loud shots reverberated through the thick air. Everyone, in shower sandals and underpants, immediately grabbed their FAMAS and jumped into action. We ran down the hill in the direction from which the shots had come. Halfway there, in a small clearing illuminated by the moon, we saw the sentry standing silently, still with his FAMAS by his side. We didn't know if he was shooting at food, fired the initial rounds as a warning, or if we'd just been drawn into an ambush.

Escaping the Amazon

As soon as we made eye contact with him and before anyone could even register what was happening, he placed his FAMAS barrel into his mouth and pulled the trigger. His eyes widened, and the back of his skull erupted in a spray of blood and flesh. He dropped ungracefully to the ground. Everyone stood motionless—like his limp body—in shock. I stared at him lying on the ground, his eyes still open and filled with horror. Without asking, I understood all too well what pains led this young man to take his life. For the death of a stepson of France, his coffin would be draped by the tricolor he swore to defend and perhaps that of his home country. The body of this boy, who never existed, if unclaimed, might be buried near the regiment, or in a sleepy forgotten plot in France, under the name given him by the Legion.

All life is a gift and not to be discarded, yet suicide is never a simple matter. On September 11, 2001, I watched office workers jump from the flaming Twin Towers in New York City. I couldn't reconcile whether their action was suicide or simply a desperate attempt to escape the excruciating pain. I wondered if this poor legionnaire's spirit would be left to wander the jungle, searching in vain for his *section*, or to seek justice and redemption.

JOB

An elite soldier: you will train vigorously, you will maintain your weapon as if it were your most precious possession, you will keep your body in the peak of condition, always fit.

Code d'honneur

We were usually oblivious as to how easy it was to die in the Amazon. In combat, the jungle itself was often a bigger threat than the enemy. On occasion legionnaires went missing, only to be discovered drowned, with their foot stuck between twisted roots.

On one of Connor's missions, he was tasked with clearing a fallen tree, which was impeding the *section*'s footpath. I didn't believe in karma, but I was sometimes tempted to. As Connor stood on the trunk and assessed the situation, a nasty sergeant who seated a grudge against Connor, shoved him off the tree and into a sharp, woody bush.

"*Depeché tois!*" the sergeant shouted and jumped over the stump, showing Connor how to get the job done. But a second later, he crumbled into a ball and gripped his ankle, bawling like a baby. Connor got up to help and saw a pitch-black snake with a distinct white stripe across the side of its head slither away. Connor quickly called a medic to inspect the sergeant's leg. Just above his boot was a clear bite mark. After hearing the description of the serpent, the medic immediately got on the satellite phone.

"Medivac, come in. We need immediate assistance."

Not able to hear the other side of the call, Connor and the rest of the *section* were left guessing about what happened next.

"Roger that," the medic ended the call.

"How long before the chopper arrives? Should we start clearing the area for evacuation?" Connor asked, ever eager to aid his tormentor.

"They're not coming," the medic mumbled.

"Say that again?" the sergeant protested as he sat up.

"They are not fucking coming," the medic stressed for all to hear. "Based on the description of the snake, if it bit you properly, you'll be dead by the time they get here. We have been ordered to carry your body out. But they did have the courtesy to suggest we make you comfortable before you die."

"Are you fucking kidding?" the sergeant pleaded in a panic. "Call those cocksuckers again. I'm a fucking sergeant. They can't just leave me to die!"

The medic seemed unperturbed by the sergeant's rant and dug in his med-pack. He pulled out a small syringe and a tiny glass vial. Like the "insane asylum scene" in a B-film, the burliest men approached the sergeant from behind and held him immobile while the medic jabbed him with a sedative. The sergeant immediately became happy and docile.

"Connor put up his hammock and make him comfortable," the medic instructed.

"Say your goodbyes and then set up your own bivouacs," the corporal added.

Around the bite, the sergeant's leg was slowly changing color, now all the way up his calf. After everyone had said their farewells, they lit a fire and opened a bottle of rum, which they had confiscated from smugglers. Not expecting to see the sergeant alive the next morning, they toasted as if already at his funeral.

"Is there a priest among us?" asked the medic nonchalantly.

Nobody responded. Finally, a quiet legionnaire raised his hand fitfully.

"I was raised protestant and studied the bible as a youth. I'm a sinner but can lead us in prayer," he said.

A Muslim legionnaire gave his respects in Arabic.

While Connor stayed up to keep the dying man company, the rest of the *section*, like the disciples in Gethsemane, fell asleep. But when light broke through the jungle ceiling the next morning, they were all amazed to see that the sergeant had somehow survived.

"Lazarus! It's a bloody miracle!" the medic announced as he checked his vitals.

A stretcher was made from tree branches and a legionnaire's combats, and the sergeant was carried back to base, to the front door of the first responders.

The heart of the Amazon was still largely unexplored, and new tropical diseases were often discovered when troops returned from the interior ill or dying. Though we were inoculated for some maladies, I'd met legionnaires who contracted malaria, dengue, or yellow fever. The latest French medicine helped, but there was no sure way to avoid those parasitic ailments other than to pray that any one infected mosquito select a different victim.

Many an Amazon explorer had come and gone, often losing his mind, or worse. The Lost City of Z was the name given by Colonel Percy Harrison Fawcett, a mad British surveyor, to an indigenous city that he believed had existed in the jungle of the Mato Grosso region. Based on early histories of South America and his own explorations, Fawcett theorized that a sophisticated civilization once existed and that isolated ruins may have survived as proof. He died under mysterious circumstances, in search of it.

After one mission into the interior, I noticed a ringing in my ear and diminished hearing. It was initially a blessing to not hear our corporal barking in my face. I suspected that I'd damaged my hearing during a firing exercise a few days prior, yet I had worn ear plugs the whole time. I then considered that river water might have permeated my middle ear. By the end of the day, I could barely hear anything.

The next morning I slept through the company-wide reveille. It wasn't long before a corporal stormed in to sort out any sleepy-heads who didn't think the rules applied to them. My roommates were packed and ready as he started shouting. But it sounded as if I was hearing him from the bottom of a swimming pool. He slapped me hard across the ear. When I lifted my head, his mouth stopped moving, and his anger subsided, for he realized that something was wrong. I turned to look at my pillow, and it was covered in blood. When I then faced the corporal, he was pointing at my ears, which were warm and wet when I touched them. Unimpressed by blood pouring from my head, I got one more blow for good measure before he stormed out. The Legion always believed that any medical ailment could be remedied with a fist.

But ten minutes later, the corporal barged back in with medic in tow, agitated that he now had to deal with this right before *rassemblement*. Despite his brutishness, the corporal mustered a genuinely concerned demeanor as he instructed me to lie back down while I was worked on.

The medic had one look inside my ear and ordered me to follow him. I wasn't instructed to simply check into the infirmary—I was told to sprint my ass there immediately! We hopped, skipped, and jumped our way across the manicured gardens and the parade ground to get there. My head felt like an overinflated balloon, as warm blood ran down my neck. The only sound I heard was my heart beating.

A French army doctor collected me at the door, with a brief exchange of words between him and the medic. I was placed on a gurney, poked and prodded, and swallowed a cocktail of pills in an attempt to stabilize my condition. During all this tumult, I was never given any shred of information as to what was actually wrong with me.

After the first week in my isolated room, I was worried about whether I'd ever hear again. My heart was no longer throbbing out of my chest, and the bleeding had ceased, yet I was still mostly deaf. The doctors, more concerned about their wives in Paris and their overseas bonuses, didn't seem to be too worried. And when I did have the nerve to ask what was ailing me, I couldn't understand what they were saying. I wasn't about to fight the system and just rode out the experience in the hopes of recuperating.

Being unharassed, rushed, or traumatized was rare in the Legion, and my stay in the hospital allowed me to indulge in undiscovered intellectual pleasures, namely literature. As a youth, surrounded by the temptations of rugby and girls, I was never drawn to reading, however, the depths of boredom illuminated even that which was previously mundane. Any mental escape or flight of fancy would do. When the real life Count Dracula, Vlad Tepes was finally imprisoned, he killed time by impaling flies on twigs. I had no idea how the American classic *A Farewell to Arms* by Ernest Hemingway ended up in my nightstand in Guyane, but I devoured it, fantasizing about fighting in the Spanish Civil War, bearing arms for ideals, and dying heroically for a just cause. Two days later, I picked up a recent tome, *Kite Runner* by Khaled Hosseini. In it, I imagined running around my family farm, challenging the protagonist Amir to a kite battle. Before I got carried away by it all, I'd be poked by an orderly and told to take my pills.

As I looked out my window, I'd see *sections* returning from mission or a course. The most raggedy bunch of men I saw were those who'd just finished the notorious Advanced Jungle Warfare Course at the *Centre d'Entertaînement à la Forêt Équatoriale* (CEFE, Jungle Training Centre). A few of them suffered from the same condition as me. But regardless of their desperate state, they were all elated to have earned the esteemed jungle badge, the ultimate test of manhood.

Escaping the Amazon

As such, my next two weeks continued, until one morning when I woke up to the most beautiful sound of a Cuckoo chirping in the distance. Like Ebenezer Scrooge on Christmas Day, I flung open my window and praised my beloved Foreign Legion for restoring my hearing. The doctor explained that I'd somehow irritated my inner ear with ear plugs, a cotton swab, or while showering and that the witches brew of bacteria in the rivers found a cozy space to colonize and multiply, leading to swelling and closure of the ear canal. I was lucky to have received medical attention early on. In true Legion fashion, once I was barely able to walk unassisted, I was chucked back into my *section*.

"Did the doctors fix the boo-boo in your ear?" asked the corporal who had previously smacked me. "They better have, because you need to pack your rucksack. We're sending your ass off to the Advanced Jungle Warfare Course. Get moving!"

This course was coveted not only by every legionnaire, with only a few given the opportunity but by every other military force in the world. It was common for participants to lose a kilogram of weight per day. Rumors and fear mongering abounded as we feverishly prepared for what even the most hardened men dreaded.

I would now finally prove myself not only to my company and commander but simply to myself. To make things still sweeter, Connor would be joining me. Still recovering, I wasn't sure if I was ready for this after lying in bed for weeks, but like all things in the Legion, we learned by a baptism of fire. I'd come to Guyane for this. Go big or go home.

It was the best of times, it was the worst of times. "*Depeche toi, putain de merde!*" I heard even before our transport came to a halt. "*Sortir, sortir, sortir,*" the shouting continued. "Pushups," the finishing command that every legionnaire expected.

121

I hadn't even a chance to see where or from whom the shouting had come, but it didn't matter. We knew what to do. *Just drop your head, grit your teeth, and embrace the suck.* Fatigue and adrenaline made my arms tremble as we approached seventy pushups.

"*Soixante dix,*" we shouted in unison, some guys barely making it up.

Flash grenades were thrown around and between us. The pressing jungle heat was suffocating as we gasped for air between presses.

"*Rampez a la corde!*" the next order came as we all dropped to the ground and crawled our way to the ropes hanging on the far side of the muddy terrain.

"*Montez!*" the shouting continued.

Connor jumped up, being the first one at the ropes and ready to climb. As he situated his feet, with his back erect, a flash grenade flew over our heads and dealt a blow to his side. After the violent thud, his wind knocked out of him; he instantly fell to his stomach. That was enough of a warning for us to always start the ascent from a lying position. As I climbed, my hands blistered, as the mud made it impossible to grip the rope without slipping.

Five minutes in, and the course was already living up to expectations. *I'm alive,* I thought to myself as the sergeant barked out more orders. He came and shouted into my face that I was useless. I loved it!

Sergent Petrov was a tall, broad shouldered, blond—almost white—haired son of a bitch. His ugly mug was covered in acne scars, and his eyes carried the years of danger and death he'd experienced in the Amazon. Orwell quipped that by age fifty, everyone has the face he deserves. This guy jumped the queue by twenty years. He sported a mullet—in defiance of all military regulations—to show everyone that this was his hell hole where he called the shots.

"I love the smell of Napalm in the morning," he shouted in a thick Russian accent as he threw his hands in the air, showing us he had already lost his mind.

"It smells like...?"

"Victory!" we managed to shout from the muck we wallowed in.

"Enough bullshit. *Rassemblement!*"

Finally, I thought, as we scurried off, barely able to carry our own weight, to the formation area.

"*Depeche toi!*" *Sergent* Petrov kept at it, as we were obviously moving too slowly for his taste.

"For the next two weeks, your souls belong to me. Your rank does not matter here. All that matters is my command," Petrov dictated as he marched up and down in front of us with his hands crisply behind his back.

"Take off your rank chevrons, your name badges, and your berets. Grab one of these bucket hats with a number on and guard it with your life. From now on you are nothing but a number to me."

"There are only three rules here," *Sergent* Petrov continued. "One. The only words that you are allowed to use are *Oui, Sergent*. Two. The only time that you are allowed to stand still is when I call *rassemblement* and when you are sleeping. The rest of the time you are to be running, even if you're waiting. Three. Whenever you start or finish a task, you shout *SELVA!* Am I clear?"

"*Oui, sergent!*" we all barked in unison.

"It takes me five minutes to walk down to the river and get into a powerboat," *Sergent* Petrov continued. "Once I'm there, I'm leaving. If you, your rucksacks, your FAMAS, or food and water are not in the boats, you'll be swimming after me with all your gear tied to your back. *Action!*"

"*SELVA!*"

123

Pandemonium broke out as everyone scattered about like headless chickens. Seconds later, as our senior leads established order over the mob, a few of us got smacked across the head. Each legionnaire was instructed to add value where they saw fit.

"Darklay, *sac a dos avec les eaux. Depeche toi!*"

I threw my gear and two twenty-liter water bottles onto my back.

"Connor, *la bouffe.*"

"*Oui, caporal!*" Connor responded as he ran up beside me. "This is fucking awesome," Connor whispered in my ear as he grabbed his rucksack and a case of rations.

"You still seriously love this shit or just kind of like it?" I squeaked.

"Love it!"

Connor was no mere mortal.

We panted and ran down the hill towards our powerboat. The rest of our section behind us was also stumbling down the slope, overloaded with sacks, crates, and whatever else we needed for the next two weeks. We looked like a colony of worker ants carrying enormous loads, chasing after their queen who was about to relocate the nest. Except that our queen was far from comforting or regal, and our new nest was a muddy pit in the middle of the Amazon.

Sergent Petrov started the countdown.

"*Dix, neuf, huit.*"

Bodies and cargo were thrown into the two metal boats, each with a sturdy 100 hp outboard motor. The side of the vessels were painted in camouflage drab. Inside were rows of wooden benches for us to sit two-by-two. Connor was then sent back up to fetch more water.

"*Sept, six, cinq.*"

I looked back to see who we were missing, only to see Connor running towards us and then falling head first ten meters from the boat. Putting his body on the line, he focused on keeping the water out of harm's way and used his face to break the fall of his 110-kilogram body.

"*Quatre, trois, deux...*"

As if Connor had planned the move the whole time, he tucked in, rolled twice, and then popped back onto his feet with a mouth full of mud. He tossed me his load and hurled his body towards the boat.

"*Un! Action!*" *Sergent* Petrov howled, not concerned whether Connor would make the leap or not.

Our powerboat sped off as Connor barely got an arm inside it. As he was dragged through the water, his hands began slipping down the side, taking his body closer to the propeller, now at full speed. I leant over, grabbed the back of his combats, and pulled him safely in. Drenched, Connor lay there panting.

"What the fuck?" he grumbled as he sat up, grasped his FAMAS, and took his place.

The powerboats took us across the river and several kilometers up the bank where we slowed to approach a jetty protruding out of the jungle. As we moored the vessel, the mad rush began again, this time with better coordination than we had previously achieved. Corporals directed the flow as we unloaded. *Sergent* Petrov watched from a distance. As soon as he saw an idle body, he flew into a rage.

"Number 4, are you on fucking holiday? Drop and give me pushups until your *camarades* are finished. Enjoy your break. *Putain de merde*," he yelled as a vein in his neck pulsated.

Hushed sniggers ran through the *section* since Number 4 was a sergeant.

125

"*Rassemblement!*" Petrov howled as if the last rations box to touch the floor was his cue.

We scrambled twenty meters up a steep muddy hill to the only flat piece of ground we would see for the next two weeks, measuring roughly 20x20 m. To the right was an obstacle course with an array of ropes between different structures. To the back left were muddy technical obstacles. The former was the individual course, which every man would have to complete. The latter was the team obstacle course. The record time for it was 45 minutes and had stood for several years. The slowest time was 7 hours by a visiting US Marine Corps contingent.

A way off was a steep, muddy, 45-degree footpath that fed into the jungle. It had been eroded by years of rain showers and boots, making it look more like a kid's slip-and-slide from hell. It should have been designated instead as a feet-and-both-arms-path.

"How the fuck are we going to get our gear up that?" I muttered to Connor.

"I bet we're about to find out now," he replied with a smirk.

Sergent Petrov began shouting, and within seconds we were all standing in formation, which was good progress from the earlier mayhem. That opinion was short-lived.

"You thought this would be a tropical holiday, with cocktails and free whores? When I say *rassemblement*, I mean immediately!"

The entire *section* groaned as everyone braced for what was coming.

"Now grab you FAMAS, straighten your arms, and hold it in front of you."

With the FAMAS weighing less than four kilograms, it seemed a reasonable task, however, after 45 seconds, our arms were aflame. After a minute, a sharp pain shot into my shoulders. Another minute and our backs trembled as we compensated by leaning backwards. We grimaced and groaned, but nobody wanted to be the first to dip even an inch. By minute three, my mind filtered out irrelevant stimuli, including *Sergent* Petrov's commands. I focused solely on my shoulders, which now felt like they were holding up an entire armored personnel carrier.

"Pushups *kurwa!*" *Sergent* Petrov bellowed after my arms eventually collapsed.

I dropped to the ground and gave him twenty. Now able to give my other muscles a break, the pushups never felt so good. I got up as slowly as possible to delay continuing the previous torture.

"*Rassemblement!*" he shouted, seeing that not one of us could hold up our FAMAS longer than ten seconds at a time.

By this point, my excitement dwindled, and I was now utterly spent. It was clear that my weeks in the infirmary had deteriorated my level of fitness. Yet I stood erect and controlled my breathing to hide any weakness from *Sergent* Petrov. *Poker face, Alex,* I thought to myself. *This fucker won't break you.*

"It is now 17h30. As you know the sun is gone by 18h00. You have until then to get all your gear up the hill to set up your bivouac. Take this stairway to hell. Number 12, do you know why we call it that?"

"*Je ne sais pas, sergent,*" I shouted, not knowing if that was a trick question, for we were only allowed to answer *Oui, Sergent!*

"Because every morning you'll descend here to join yours truly, Beelzebub, in hell, where I'll turn you into the best Amazon warriors the world has seen."

Despite my fatigue, his words lit a fire in me. Everyone bowed up as we basked in that short moment of paternal mentoring. We were his new project, and in *Sergent* Petrov we trusted.

"*Action!*"

"*SELVA!*" we yelled in unison.

Petrov delighted in watching everybody scramble up the stairway to hell, once again seeking out any laggards. He had an eye like a hawk and moved like a snake through the jungle. As soon as we thought he was out of sight and we could catch a quick breath, he'd magically appear like a leprechaun shouting "pushups, *kurwa!*" We would drop to the mud to slog out a few pathetic pushups, regretting cutting any corners under his charge.

Only years later did I remember an odd image that burned itself into my memory. While the mass of legionnaires was scrambling about eating mud, as *Sergent* Petrov basked in delight, across the bank stood an indigenous elder. He watched, with curiosity, while the crazy mess of so called civilized men ran around like buffoons, destroying their own members. Between breaths, I caught a glimpse of the indigenous man walking back into the forest, either confused or saddened.

By 18h00, our bivouac was up. I expected to be robbed of all but a few hours of sleep, but that wasn't quite the case. The Legion wasn't going soft on us, though. In the Amazon, the sun rose and set at around 06h00 and 18h00 respectively, with the forest being dangerous even at high noon, and moving around after dark was suicidal. Only in the Legion was a seven-day, twelve-hour work schedule considered an undeserved treat. Like warring Afghan tribes who ceased fighting during Ramadan, we organized our schedule around Mother Nature.

But within every workday, *Sergent* Petrov packed in more bullshit than a Texas feed-lot. Every movement or task was conducted *pas gymnastique*, and if we ran out of work, he'd give us any other sadistic exercise that came to mind, from duck walks—walking on our haunches—to holding out our FAMAS while squatting in a sitting-chair position.

Each evening we descended to the riverbed for our daily shower. Naked male flesh waded in the murky waters, with snapping turtles waiting to chomp off any juicy appendages. But our biggest nightmare was the Candiru fish, a tiny barbed predator that is attracted to urine and embeds itself deep into the penis. Removal without surgery was impossible, and the pain was excruciating enough for men to sever their member or commit suicide. Piranhas also trolled the waters. We had to wash our combats, boots, socks, and underpants to be clean and fresh the next morning. We would then walk up the muddy stairway to hell back to camp—getting filthy in the process—set up our bivouac, and hit the hammock for the night. Some men would chat and laugh until a random sergeant lost his temper, chucked a filthy boot, and shouted a string of expletives through the jungle—and then silence. For 06h00 *rassemblement*, our bivouac needed to be packed up, our rations divided out for the day, boots polished, and all of us at attention.

Sergent Petrov would briskly inspect our uniforms. Even in the midst of unending mud, sweat, and lack of soap, parade standards had to be kept for the duration of the course. And as expected, there was always a wrinkle or crooked collar that warranted collective punishment, which left us and our clean combats covered again in muddy saliva and sweat.

We would then be instructed on how to navigate one or more group obstacles, to prepare us for the timed test toward the end of the second week. We'd be briefed on the day's schedule, and then be given tasks, each interspersed with more running to and fro, until sundown.

Each lesson was based on one of three major principles, jungle warfare, jungle survival, and teamwork. The itinerary built upon itself, leading to our final days when we would be dropped into the rain forest for a special assignment.

During one survival module, we hiked up to a makeshift amphitheater where we were introduced to a local Amazonian who would teach us everything we needed to know about the jungle. We had no idea where the Legion found this ascetic hermit, for he was neither Guyanese nor Brazilian. The rainforest was his home. His passport must have read Jungleistan. He took great pride in teaching us the ways of his unique homeland. Every eye glazed over. My section was either wholly enthralled by this holy man, or just happy to get away from *Sergent* Petrov for a few hours.

His instruction was clearer and more informative than anything I had heard in the Legion. We learned how to make rope, how to use it to climb trees, and how to create a backpack—all with a humble palm leaf. We were taught where to find our food and how to prepare it. Every type of fruit, egg, bug, or plant that could be eaten was arranged in front of us, to inspect, feel—and taste We were also shown a variety of common poisonous plants, which could be used either as a weapon or medicine, depending on the dosage.

How many people had died to confirm this knowledge? I wondered, captivated by the Amazonian's comprehension. The most interesting noxious plant was Angel's Trumpet, a flower used by shaman to produce lucid dreams, which in turn are analyzed to diagnose diseases or predict misfortune. If used in excessive doses, it could also cause death. Communicating with Jimi Hendrix seemed a good excuse to get high, and depending on the hit, it may have been an experience to die for.

Our day passed in the blink of an eye and, as the jungle started to cool, evening approached, and with it, *Sergent* Petrov.

"*Rassemblement!*" he announced.

We all scampered to our places as quickly as possible.

SMACK! I heard the sound of pain behind, and turned to see Connor in the fetal position after dropping a load of jerry cans filled with water. Like a quivering dog, he got up and darted in the opposite direction to the pain.

"Why aren't you ladies running?" *Sergent* Petrov shouted as he reached for the next unfortunate victim. Instantly the whole *section* began trotting, not wanting to receive another blow of our own. Being exceptionally tall made Connor the favorite whipping boy, to be made an example of, or to be used as the *section*'s mule. True to form, he took it all in his stride and never muttered a complaint.

We were soon back at our bush camp to set up and get a night's reprieve from *Sergent* Petrov. Being mindful of the morrow's timed obstacle course, we all wanted to rest as much as possible to prove who was the best commando amongst us.

"See you ladies in the morning," *Sergent* Petrov boded.

"*SELVA!*"

Every man was already up before reveille, packed and ready to go, for this was marathon day. I marveled at the variety of typical, unique, or bizarre preparation rituals. Some carb-loaded on double rations. Others engaged in yoga stretches. I simply looked to the heavens and asked my dad in heaven to watch over us, and for Jannie to ride shotgun with me on this drive. One could slice the anxiety in the air with a machete. Like Greek Olympians before the competition, we would have devoured raw testicles if given the opportunity.

As usual, we descended the stairway to hell to meet our puppet master at 06h00 sharp. I felt like an ancient gladiator walking through underground corridors towards the *mano-a-mano* battle. Everyone was silent. I thought back to *instruction* when our attitude was akin to that of forced conscripts. Yet here in the heart of the Amazon, reserved for the crème of each *section*, every man was high on testosterone and fanatically competitive. The first event was an individual effort, and teamwork was out the window.

We dropped our gear on the ground, our FAMAS beside it, and stood at attention in perfect rectangular formation. Nobody moved a muscle, for we were ready. The faint sound of outboard motors could be heard in the distance. *Sergent* Petrov was approaching.

"Morning, ladies," Petrov said crisply, announcing his presence, standing at the helm of his powerboat, *a la* General McArthur. "Today we will crown the best of the best. Are you ready?"

"*SELVA!*"

"Shall we get started? Time for a little warm-up. All of you, down and give me fifty. I don't want any of you straining a pectoral while pulling yourself over the obstacles. If you're going to fail, let's get it out of the way now."

By the time we were limp from exhaustion, *Sergent* Petrov finally declared that we were limber enough to take on the tests.

"Number 1. You're up."

"*SELVA!*"

We systematically queued in order behind a pole planted at the foot of the muddy formation ground, which served as the start and finish line.

"*Action!*" Petrov shouted as the first participant shuffled away to confront the first of three earthen obstacles. They were interspersed with wet ones, which we had to slog or swim through. Our Olympic level swimmer, Connor, had a distinct advantage over us mere mortals in that respect. A few guys finished before I was up. They stomped their way to the finish pole to stop the clock and then flopped to the ground entirely spent.

"Number 12. *Trois, duex, un...*" *Sergent* Petrov counted down as it was now my turn. "*Go, go, go, kurwa!*"

I flung my body at the obstacles, with utter disregard for my physical wellbeing. I chose to jump off of a three-meter wall and save a few seconds rather than climb down safely. My heart was in my throat, and my lungs tasted of blood, yet every fiber in my body was fighting my arch opponent: the clock. As I lugged myself over the final obstacle, I could see the finish pole and a few more blokes lying on the floor beside it. As soon as my feet finally touched the ground, I started sprinting. My boots felt like 50-kilogram dumbbells. It took every last morsel of energy to finish, as I had absolutely nothing more to give. *Leave it all in the ring. Knock out your opponent. Never let it go to the judges...*

"*SELVA!*" I shouted and collapsed with my fellow legionnaires.

After a real nail biter, it was time to crown our champion.

"Number 7. Come up for your handshake. You are today's fastest finisher," *Sergent* Petrov announced nonchalantly for the Legion's typical anticlimactic celebration of personal achievement.

As our own mad Aussie, Connor, turned towards Petrov, a slight smirk of pride emerged from the corner of his mouth. *Once a warrior, always a warrior.* Whether it was in a clear chlorinated swimming pool or muddy cesspit, that man earned every bit of that win.

"Why the hell are you ladies just standing around?" *Sergent* Petrov asked, returning to his customary demeanor. "Drop and give me fifty." A tiger never changes his stripes.

"*SELVA!*" we shouted, though we could barely budge, much less do a pushup.

"Thought you were done for the day, huh? Next, we move onto the team obstacle course."

"The fuck? I thought that shit was for tomorrow?" I grunted to Connor as we continued with the pushups. "Evil bastard."

"Each obstacle will be a gauntlet of sorts. It will take team work and coordination to overcome them," *Sergent* Petrov continued. "I'll assess from all angles, seeing who has the best leadership, who has the best listening skills, and who has the best judgment in attacking each obstruction. *Action!*"

We struggled up, regrouped, and immediately took on the course. It was natural for the corporals and sergeants to take the lead. We systematically worked through the initial obstacles with relative confidence, pacing, and logic. With nothing to compare our performance to, we had no idea that we were progressing too cautiously and slowly.

Nonetheless, I finally appreciated how the Legion's clunky ways of doing the most menial tasks ultimately worked in an elaborate chess game where pawns labored towards a common goal. Hanging upside down from a three-meter high obstacle and carefully passing a massive log—not allowed to touch the ground—between my arms, punctuated our role in the Legion. We had to trust our superiors' logic and best practices to get the log from one side to the next. Our leads had to trust that we would risk our lives to keep it from dropping. Our calling was to do stupid things, like lug five guys over a muddy wall and into quicksand, but there was somehow an exceptional beauty to it.

We ended up finishing the team obstacle course ten minutes slower than the expected time, which none of us took lightly. *Sergent* Petrov sure as hell let us know that we had failed.

"Did you faggots drop your lipstick in the mud? Give me fifty! I would have completed that course on my own quicker than you could pull up your panties."

After the pushups, we moved halfway up the stairway to hell for the third and, presumably, final test. It was another individual course made up of a web of ropes between poles and trees. If anything so much as a bootlace touched the ground, we'd have to start over. The gauntlet progressed from having to walk a tightrope between trees—which usually became a sloth-crawl while clinging upside down—to swinging like Tarzan from one hanging rope to another. Connor's height, which gave him an advantage during the muddy course, wasn't going to help him one bit in this scenario. One Czech ex-gymnast left us all in the dust as he crossed the ropes like a stone skipping across a river.

After all three courses, our bodies were demolished. Every muscle was in agony, and our hands were cramped and bleeding. Grit had chaffed our groins and armpits. Our joints were tender and swollen. But as feared, our day had only just begun.

"*Rassemblement!*" *Sergent* Petrov trumpeted.

Exhausted, we all sluggishly moved back down the stairway to hell.

"Grab your FAMAS, and your machete and get into a powerboat. Put your faces down on the floor and don't look up," he instructed us. We obeyed. As we kissed the steel hull, the engines roared, and we plowed down the river. Where to, nobody knew. After half an hour, the boats slowed and pulled into a break in the jungle where we were ordered to disembark.

"Two kilometers down this river on the other side is safety. *Sergent* Petrov explained as we tried to find our bearings. "You'll have to survive three days in the Amazon with nothing more than what you have on you: your FAMAS, jungle knife, and your newfound knowledge. In that time, you must somehow build by hand two rafts secure enough to carry the entire *section* to safety,"

"Shit," I muttered to Connor.

"You will erect a shelter. You'll craft every conceivable trap, and you'll have to eat whatever you catch. See you ladies in three days... Or not!" *Sergent* Petrov jeered as he lit a Gauloises, climbed back onto his powerboat, and sped off.

After that morning hell, our bodies were screaming for food. We were already dehydrated and hadn't eaten for a day. Connor started chopping a pathway into the jungle. Our first priority was to build a secure shelter while we still had daylight. It also needed to be raised off the ground if we wanted to avoid the creepy-crawlies and other unpleasant predators that inhabited the jungle floor. Instinctively, half of my *section* set out and found and prepared a sufficiently large clearing, while the other half started chopping down branches to use for the shelter's base. We lifted the base a meter above the floor, but it was only large enough for a dozen romantic couples to snuggle closely together. A roof support was made out of the same branches and then covered with layers of palm leaves, such that rain would run off without seeping through and onto us. We then split into three groups according to our best skills. One hunted animals and searched for water, the other set the traps, and the third began constructing the rafts.

Nightfall came quite quickly, and our hunting and gathering efforts of the first day were short lived, leaving us still starving. We started a fire from the scarce dry tinder we could find and fed it with damp wood. It provided some warmth for those huddled around it, and the smoke deterred predators. Despite our exhaustion, nobody slept much. At the crack of dawn, we slowly emerged from our holes, took a breath, and immediately got back to work without muttering a word.

Connor and I were on raft duty that morning and were chopping down a large tree earmarked to form part of the base. I knotted the deck to the foundation with the rope I had made out of palm leaves. Soon, we barely had enough energy to swing our tools. In the jungle, inattention, fatigue, and sloppy practices can be deadly.

"Ouch, bastard!" Connor snapped as he jumped up. "A bug bloody bit me."

I rushed to save my *camarade* in distress.

"Look on the bright side. If you die from the bite, we can all eat your carcass and not worry about hunting," I joked.

"That's only if we're stranded in the Blue Amazon. When I promised the boys, I'd make them a good Australian dish, that wasn't what I had in mind."

As I got closer, I saw that Connor had perched his back against an ants nest.

"Shit," I said, "if you were injured and immobilized, those bastards would have devoured you in minutes."

Without saying another word, both of us picked up long twigs and pressed them into the nest. We extracted an angry mob of ants and instinctively lifted them to our mouths.

"They eat us, or we eat them. Law of the jungle."

"Nice lemon taste," replied Connor with a smile and ant guts between his teeth.

Others heard of our special snack and came with sticks in hand. Ten additional calories didn't amount to the difference between life and death, but the psychological boost was enormous. We got back to chopping down the tree with a bit more pep in our step.

Later we discovered some heart of palm, which was bountiful and rich in potassium, iron, and B vitamins. We weren't about to expend tremendous amounts of energy chasing animals to eat. If John the Baptist could survive indefinitely on honey and locusts, then we could surely get through three days with little or no nourishment. Finding food often contributed to death in the wild, as most victims didn't know that the human body could survive weeks without it. Water was a different matter.

By nighttime, morale began to drop, and men got into petty squabbles. We all moved like zombies. The rafts were only halfway completed, and the work was slowing. But our final sunrise gave us a second wind, and we beavered away finishing our escape vehicles. Our sole motivation was not being stuck in the jungle another miserable night. By midday, we had two rafts strong enough to withstand the ravenous river and carry us all to safety. We split into two groups to make sure the weight in each was even. We also divided the best swimmers, since they'd physically guide the boats and serve as lifeguards.

Like Lewis and Ed in *Deliverance*, we were finally on our escape rafts, battling the rapids towards safety. The journey was going as planned until we were a kilometer down and realized that we had to quickly navigate across the river to avoid a forking distributary.

"Connor, in!" a sergent commanded.

Connor didn't hesitate, as I believe he preferred being in the water than on the raft. He'd happily swim alone to our destination if allowed to do so.

"Darklay, in," the next instruction came, as the *sergent* realized that more horsepower was needed.

Three hundred meters later, there were six legionnaires in the rapids pushing and pulling our craft across the current. Having eaten very little for the past days, every stroke was taxing. As my glucose levels approached zero, my vision became blurry, and I felt dizzy. I was about to pass out under the water!

A kilometer later, my toes and fingertips finally made contact with the sandy bottom. We had somehow managed to push our raft across the current to the other slimy shore, our destination. We had made it, our tasks completed for the day—or so we thought. We pulled our rafts to the beach, and everyone disembarked, giving high-fives to each other in celebration. All of our other gear was awaiting us on the bank. And then we heard the distinct sound of a Peugeot P4 jeep. We looked around and saw a blond mullet-sporting driver chugging down the hill towards us.

"Petrov coming to shake all our hands?" I asked Connor.

"*Rassemblement!*" *Sergent* Petrov announced with a stern voice, as his vehicle skidded to a halt.

"What the hell, now?" Connor asked in a whisper.

"Number 5 and 13, you have both just been injured. Lie on your backs. The rest of you, grab your gear. Find a way to get your *camarades* to safety."

"I thought we were done!" I bitched.

We placed the poles over the stomachs of the "injured" and a pair of combats under their back and heads for support while carrying them. With our palm leaf rope, we tied them from head to toe to the pole, like pigs on a spit.

"Safety lies through that gap," *Sergent* Petrov continued as he pointed to a footpath that would take us back deep into the surrounding Amazon.

"*Action, kurwa!*"

We rotated amongst four porters who transported the injured over, under, and around various impossible obstacles. We couldn't have them under water for more than a few seconds, to avoid drowning the injured. Nor could we drop them and possibly break their necks. Those poor devils were immobilized and entirely at our mercy. There wasn't a minute that passed without them cursing us for banging their heads against another tree or dragging them across a thorny bush.

"If you were really injured, you assholes wouldn't be complaining. So just enjoy the ride," we joked.

After an hour of painstakingly maneuvering through the thicket, we finally emerged onto a dirt road, which we knew lead to where all our misery had started. This was the beginning of the end of the toughest two weeks of my life. Ironically, with relief in sight and our bodies dragging, finishing the course felt bittersweet—by now we enjoyed the suffering and wanted even more. As we picked up our trot, our spirits rose, and we all broke out into song, like a band of brothers running to victory. *This is why I joined the toughest army in the world!*

We came around the last corner and could see *Sergent* Petrov standing on the sandy parade square waiting for us, the first time we ever saw him smiling.

"*Un* [stomp, stomp, stomp], *deux* [stomp, stomp, stomp], *trois* [stomp, stomp, stomp], *quatre* [stomp, stomp, stomp], *un-deux-trois-quatre, un-deux-trois-quatre, SELVA!*" chanting in cadence as we approached.

"State your business," *Sergent* Petrov demanded.

"*Sergent* Petrov, mission accomplished," our lead replied.

Thirty minutes later, we were all dressed in neatly pressed combats and at attention.

Sergent Petrov made his way through the entire *section*, personally shaking hands with every single one of us, pinning our jungle badge to our chest and saluting us.

"Darklay, you are now a jungle commando. Well done."

"*SELVA!*" I shouted, loud enough for my family in Johannesburg to hear.

ASH WEDNESDAY

The edge of a colossal jungle, so dark-green as to be almost black, fringed with white surf, ran straight, like a ruled line, far, far, away along a blue sea whose glitter was blurred by a creeping mist.

Joseph Conrad, *Heart of Darkness*

After completing the jungle course, everything in the REI seemed to stop, and I found myself slipping into bad habits. Depression, too, lurked close by, a lion waiting to pounce. *Cafard* afflicts a legionnaire instantly, or it consumes him slowly, like a worm that never dies, and a fire that isn't quenched. I suffered from irrational fears and felt the Amazon creeping in ever closer to me. I'd been doing little for the past week aside from boozing and passing out in a hammock. During an uncomfortable withdrawal, I finally came face to face with the company's impeccably dressed colonel, who told me that he would need me to participate in unique missions that would enhance the prestige of the REI. "You will be somebody in administration before long," he added.

Every legionnaire in the Legion contemplates desertion, either once or many times. Like an uncommitted husband who enters into marriage with a prenuptial agreement, I too wanted to have a plan in place should the fateful moment come when I chose to escape. The first step would be to leave French territory. Crossing into neighboring Suriname was an option, but without a passport, this seemed a hopeless wish likely to get me imprisoned in a neighboring country, dragged back to a French prison, or both. I would try a dry run first, to see if I could successfully cross the frontier, and then follow up with something more elaborate. I had no idea that this would be one of my many close calls with death, and not at the hands of the Amazon or the Legion. It seems I would meet Conrad's elusive madman from the jungle, Kurtz, after all.

In time, I came to learn that it was common practice for the smarter legionnaires to jump the border into Suriname for the holidays and return to the REI. No questions were asked if there was no apparent harm done. It was December, and we had graciously been given leave. As other legionnaires spent their pay on local prostitutes, I abstained and squirreled it away.

Luckily for the Guyanese, France had long ago built roads, which one could take to the border. The four-hour drive from the regiment to Suriname started in a jam-packed local taxi. Connor and I now departed with a caravan of porters into the jungle, where Marley's steamboat that I was to captain lay. Fog soon enveloped us, as we drove deeper into the vast green ocean.

The taxi continued its lurch towards the border, or so it felt. We sat right in the back with our rucksacks on our laps. Connor's knees were near his chin. My shoulder was jammed into his armpit and the rather voluptuous thigh of the mixed-race woman next to me, who crushed my left hip with her basket of produce and a caged chicken. We suffered in silence for the entire four-hour journey, suffocating in the heat, and were only mildly distracted by the drumming salsa music emanating from the worn-out front speakers. I imagined how long such a sojourn would have taken by foot, slogging through the jungle. Our old map indicated that the road we were on was paved, though it felt as if we were competing in the Dakar Rally.

We finally arrived at the border town of Saint Laurent, and everyone piled out of the taxi like a spring-loaded Jack-in-the-box. My senses were acutely active because I knew we were going to try to illegally cross the frontier without any passports. It was my first time, but it would not be my last.

Escaping the Amazon

I took in the view of the massive Moroni River. With a breadth of two kilometers, one could only just see Suriname on the horizon. The sun was already high in the sky, lighting up the lush emerald forest that framed the riverbanks on either side. The brown water was littered with hundreds of pirogues, small wooden dugout canoes with canopies, lazily transporting travelers to and fro. There was a sharp smell of fuel, generously dispensed into the air and the river by the outboard engines. A blend of languages filled my ears at the ferry terminal, characterizing the fusion of language and culture across borders. The French sounded Dutch, and the Dutch sounded French, and within that gray area, no one really belonged to one side or the other. This was the one place where the New World met the ancient one, where Europe met Africa, then backed up into Asia—with cultural permutations of all kinds. I'd been through South African shanty towns that would make one's hair stand on end, but this gumbo of vice, fast money, and degeneracy had no comparison. There was a thrilling vibrancy that arose from the bustle of pirogue pilots, officials, traders, and the occasional legionnaires like us—a backdrop for any Indiana Jones film. I took it in, momentarily forgetting my fear. As it was formerly a Dutch colony, it was relatively easy for me to communicate with most locals.

Others in our barracks who'd made this trip before advised us to just bypass the obligatory customs office stop, so as to avoid having our non-existent passports stamped. We followed that plan, and nobody halted us or asked questions. We walked down to the water's edge and casually tossed our rucksacks into the first available pirogue. We paid the ferryman—he may as well have been the mythical Charon—a scrawny, sun-scorched local, who then started the engine and shipped us to the opposite bank of the mighty river Styx. As planned, we neglected to check in on the Surinamese side and strolled past the beach towards the road to find a taxi.

145

The next three hours were absolutely treacherous as our subsequent jammed taxi flew over the worst roads I'd ever seen, at breakneck speed, with cars coming head-on every minute. For the second time since joining the Legion, I thought I was going to die, and couldn't believe the ease with which the locals took the madness in their stride. The driver didn't skip a beat as the wheels lifted off the ground when we managed a sharp corner, nor did anyone care much when we were doused with a dust cloud after someone opened the window to spit. It was just another day at the office.

Halfway to Paramaribo, the beachside capital, the taxi slowed to a stop at a police checkpoint. Without papers, we were grossly on the wrong side of the law. I couldn't fathom how the police would react, or how we were going to evade this stubborn detail. The screeching music added to my sudden terror. *I know what I'm doing*, Connor said to me with his eyes. He was as chilled as a mini-mart slushy. I followed his fatherly lead. An overweight officer smelling of curry singled us out and sidled up to our window, his stomach oozing through the gaps between his shirt buttons.

"*Paspoorten*," he grunted, without looking at us. He scribbled on a notepad, chewing laboriously on tobacco. Connor elbowed me. Trembling, I reached for my wallet and handed the policeman a crisp fifty-Euro note. His eyes then made contact with us. Though it was a mere second, it felt like an eternity. "No, you wait," he said, just as we were about to jump out and attempt a mad dash into the jungle. But he merely examined the thickness of the notes to make sure they weren't counterfeit. He returned a leer of disgust, that of a free brown man no longer under the yoke of European colonization. Connor did the same, just for good measure, and without a word, the official took the cash, spat, and waved us through. We could only pray that the controller, with our small fortune, would buy his wife a lovely dress, and take his kids out for a meal that night.

We quickly learned that it was easy to work the justice system in Suriname. Every man had his price, and with our Euros, we could easily make our way around without too much hassle. In a country where the president had been convicted in the Netherlands of drug trafficking and was wanted by Interpol, we thought we needn't take its judicial structure seriously. This turned out to be a severe underestimation on my part.

I hadn't always been a rule breaker, and often thought myself as less of a man because I too frequently opted out of unethical activities in which my peers gleefully participated. I began loathing my moral compass, and being in the Legion tested my resolve. Every passing day brought me face to face with wrenching temptations. I now understood that on the sharp end of life, it is often a matter of choosing the lesser of two evils. Who should we shoot first, the drug dealer or the human trafficker? If we confiscate their money and donate it to an orphanage, would it really be stealing? I put such ponderings aside…for a time.

And after "some time," the prince of darkness reared his ugly head. The same border official had established business with the Legion by allowing passport free passage. We set aside extra cash in order to tip him on our way back to ensure a clear route for next time.

Entering into Paramaribo, one could see the Dutch influence in the architecture, which now bore the scars of war and neglect. The streets were polluted and dirty, and white colonial buildings had small market stalls beneath them with elderly women selling local fare. A chaotic amalgamation of cars, taxis, and mopeds filled the thoroughfares with horns, which were used as a means of indication rather than as a warning. Our ride came to a halt just outside a bazaar, permeated with various rich scents of warm oily street food. We grabbed a bite and made our way towards the local Legion crash pad.

We soon arrived at the Guesthouse ZIN. It had a crystal blue swimming pool, billiards, and plenty of Kronenbourg, which made it the perfect place for us to blow off steam. Above the bar were several overlooking rooms, which were usually set aside for legionnaires, allowing us a quick walk for more alcohol and a short distance to be carried home after passing out in the evening. The main attraction, of course, was the scores of sunbathing European tourists.

A legionnaire's holiday was rather basic: sun, booze, and woman. More often than not, legionnaires skipped the arduous task of courting woman at the pool and simply ordered prostitutes from the strip dive down the street. Like me, Connor also avoided prostitutes, so we often found ourselves playing pool together when the others went off to satiate their urges. Our time away from the flesh-seeking hordes allowed us to get to know the locals at the bar more personally. We eventually befriended a local woman, Anna, through her social and coquettish daughter who had taken an innocent liking to one of our *camarades*.

Consequently, she and Anna loved inviting us to their home for dinner, since the daughter knew that the way to a man's heart was through his stomach. Connor and I, never able to turn away a free meal, ended up being chaperones-in-tow to these succulent feasts. Both Anna and her daughter were Dutch-speaking, of Indian descent, and had never ventured outside of the country. In due course, we were regularly sharing dinner around Anna's table.

Anna always greeted us at her door with a bright smile, her robust figure filling the space within the door frame. She was a mountain of a woman, a heap of motherly nurturing, stacked high and wide. When she held out her arms, the only way we could pass was by means of an enveloping hug. I had to stoop to reach around her, and Connor had to double over. She sported bright colors, which stood out boldly against her mocha-colored skin. Though always laughing, she was not afraid to smack us into line when we misbehaved.

One night, we arrived at her home just before sunset. We passed through her door and continued through a small hallway of old family portraits hanging on the wall. The television in the living room blared the evening news to anyone who cared to listen. The cat, spread out on a plastic chair, seemed the only one vaguely interested.

"Come, come sit down. Tonight I baked fresh bread and made a special chicken biryani for my darling boys." Anna loved cooking Afro-Indian cuisine, a sumptuous fusion of flavors and spices from around the world. The dining room was next to the kitchen, and the aroma of aniseed and cumin toasting in coconut oil drew us eagerly to our seats. In my haste, I almost knocked over her glass Ganesh statue behind me, which was perpetually burning incense, a reminder of Krishna's providence over us visiting Europeans. Anna didn't drink, but her meals were always accompanied by copious amounts of wine and lager, solely for our sake. We enjoyed feasting, and talking raucously for hours, with curry running down our forearms. Endearingly, Anna enjoyed having strong young men around to spoil and look after, perhaps as the surrogate sons she never had. Her daughter was now an adult, and Anna's ex-husband had either died or abandoned the family. Such a topic was never discussed, like a dog that didn't bark. And though we were fierce fighters who would have protected Anna and her home with our lives, we were barely grownup lads who looked tougher than we really were.

"Okay," she eventually said, indicating that she was about to chuck us out. "You boys go down to the casino. They have a cabaret tonight." Anna stood up and magically gathered all the empty plates on one arm. She waddled towards the kitchen, turned on her heel, and pointed a finger at us. "You better behave. You are illegal squatters in our country. I love you, but the police won't." She repeated that phrase every time we visited, but we always brushed it off. We kissed her goodbye, thanked her profusely, and left her to the dishes. Her lovely daughter, now our little sister, joined us and we spilled out onto the alleyways to walk to the casino nearby.

The rest of the holiday was a blend of partying to the early hours of the morning and then baking away our hangovers in the sun by the pool during the day. This continued all the way to the lead up of New Year's Eve, expected to be a huge night. The streets were already alive with vibrancy and color. Throngs filled every space with laughter, music, and dancing. I bumped into, danced with, and hugged Creoles, Hindustanis, Marons, Javanese, Brazilians, Guyanese, Chinese, and people who were too diverse to label.

Evenings were a kaleidoscope of events as I stumbled between bars and newly found friendships that lasted as long as a few shots of local rum. I hung my arm about every friendly chap that came around, somehow sure I knew each of them. Welcoming faces became blurry with every subsequent cocktail. One moment, I'd be atop a table downing a Parbo Bier, with all the locals cheering me on; the next, I'd be sitting somberly at the bar with a stranger, mumbling on about my spiritual funk. The rollercoaster of emotions trudged on as I downed any alcohol I could get my hands on, until suddenly, in an instant, the rest of the evening faded to black.

On the morning of New Year's Day, I woke up to the harsh sun penetrating my eyelids. I slowly became aware that I was lying on a concrete floor. I forced my eyes open, swallowed in an attempt to wet my parched throat, and sat up. A sharp pain shot up my neck, and my head was about to explode. I looked around, blinking clarity to my sight and realized that I was at an abandoned construction site. All I could hear was the remote rhythm of a party carrying on in the distance. Through my splitting headache, fragments of my memory came back to me. Connor was nowhere to be seen. I patted my hands over my torso to feel for any sharp pains. Luckily my organs were still intact. I lay back on the cement trying to cool my body. Finally, I gathered enough energy to peel myself off the floor and stagger out onto the road.

As I wandered the streets of the inner city, it was hard to believe that this was a World Heritage Site. Remnants of the previous night lay strewn everywhere. I crossed a normally busy intersection, now eerily abandoned except for a young couple walking with their arms entwined. They swayed and sang as they stepped over empty Parbo Bier cans and red firecracker ribbons. Anna told us that every year the ribbons got longer and good fortune attached the person who held the longest one. My headache returned as I remembered the thundering blasts from the night before.

I walked two more blocks towards the river and tried to recall how I ended up on that construction site. All I remembered was that at the peak of the street party, everyone suddenly packed up and left—Surinamese celebrate the New Year quietly at home with their families. Having nowhere to go, I must have wandered the streets until I passed out on the concrete. Memory blackouts are the worst horrors an alcoholic experiences, waking up and not knowing how he got there or what he did—if he had hurt anybody. If I'd been driving, I'd be looking under the vehicle for bodies and inspecting it for bloody dents.

I finally found my way back to ZIN, where Connor was sitting on the porch. A dozen legionnaires were still passed out in the lounge. Connor was awake but looked as broken as I felt. I sat down next to him, on an old worn out chair. "Happy new year, dude," Connor smiled, glugging down water. "Where did you disappear to last night?"

Veins pulsed in my head and sweat dripped down my face.

"Shit, I'm never going to get used to this heat," I said, ignoring his question.

We later awoke in those same chairs early that evening, but our hangovers hadn't faded.

"Hair of the dog," Connor said. "Let's get a drink at Jimmy's joint."

"Something tells me that we'll regret this…"

I was soon sitting at a dive bar, chatting to the manager, Jimmy, who'd been running the casino for several years. He was a fellow South African and an odd transplant in that hellish part of the world. At the front of the saloon was a small stage. A circular bar was between it and the slot machines. The roulette and poker tables filled the back of the room. A handful of old folks sat at the slots, staring blankly, hope evaporating with each coin they dropped. Most of the patrons sat around small tables near the stage. The Cabaret was about to start, and the casino was at full capacity.

Jimmy leaned over the bar, talking loudly, as usual, his gut hanging over his belt and his perennial red face betraying the fact that he had indulged in a lifetime of vice. Connor and Anna's daughter were distracted by the show, so Jimmy quickly stopped his typical rant about the lousiness of the Surinamese government and walked around the bar to join me.

153

"Darklay," he said, lowering his voice to a whisper, "can I ask you a favor?"

I leaned back slightly to avoid his fiery booze breath. "Sure man, what's up?"

"Business is tough these days, you know." He hid his hands under the table. His fingers were thick, stumpy, and trembling. He seemed to be withdrawing from drink.

"You blokes have been faithful customers and have become my mates. And, shit," he looked down, took a deep breath, and continued, "I'm working on a deal that'll free up my cash flow, but I'm short this month."

Jimmy had asked to borrow money from me before, but I always refused. I didn't trust addicts.

"Jimmy, you know I don't…"

"Just this once, please. I need to pay rent. This joint's on the brink of shutting down. It's fucking tough to run a bar in this town. Businesses have been closing down left and right."

His eyes moistened, a bead of sweat ran down his forehead, and his tremors worsened.

"You look like you need a drink, Jimmy," I said

He clenched his fists, hiding them under the table. The scantily clad dancers on the stage were weaving their way through the crowd, seductively fleecing customers of any money. I wondered how much of that went to Jimmy. After a long silence, I stared at him again. There was a look of desperate desolation on his face.

"Mate, how much?"

"Just for now, until the end of the month. I'll pay you back. I promise."

"Shit, Jimmy, stop pissing around!"

"Five hundred Euros."

"Are you out of your mind? I'm earning a legionnaire's wages. I'm not a fucking rock star!"

"I've always treated you and every other legionnaire like VIPs, with free drinks and friendly girls, but none of that comes cheap."

I stared at him like a cornered mouse.

"Come on, this bar ain't going anywhere. You know where to find me. Besides, I'd never screw my own countryman, especially in this part of the world."

To describe Jimmy as a friend would have been a lie, at best. He was primarily a peddler of debauchery. I knew nothing of his personal life other than the mindless conversations of sport, women, and war. But a latent ethical system prompted me to give this sweating, trembling mess of a man a chance. Then again, maybe I was just drunk. He poured me another shot, and I finally acquiesced.

"Fine, but if you screw me, my mates from the regiment and I'll break your legs."

"Deal! Shit, you saved my life. But I need it now. Meet me at my hotel in Room 22." I carried extra cash for emergencies. Other legionnaires did so to procure whores, drugs, and pay bribes. Either way, that slimeball Jimmy knew we had money on us.

Ten minutes later, I walked into the room, one of Jimmy's many squalid quarters that smelled of sweat and sex. I may as well have been Winston as he was led to the notorious Room 101. Hanging on the wall was a cheap oil painting of an indigenous child before a bonfire at night. I was fascinated with the sinister effect of the flame's light upon the child's face as if the artist had painted it while deep in the jungle in an opium trance. Jimmy, still trembling, interrupted me to confusingly reveal that the man I'd be meeting with was the smartest person he'd ever met, a genius of numbers and progress. "He's the real person you need to meet. I recommended you to him. He and his company run things around here. You'll go far in his administration. But just listen more than anything. You can't really talk to him…"

Before I even sat down, I knew I'd never see my cash again. I then realized why Jimmy was trembling: fear. He'd crossed me.

The room was smoky, and then entered a dozen tall, thick, pasty-white men dressed in wrinkled suits. Some took a seat while others remained standing. In the corner of the room, I could see that the bed was covered with piles of US dollars and bags of cocaine. I had the surreal feeling of having just walked into a Quentin Tarantino film set. An overweight gentleman wearing a black jacket with a high collar and large tinted glasses appeared from behind the thugs. He took a seat at the table near the window. He was smoking a cigar and spinning his Serbian-made Zastava pistol around his left index finger. Jimmy stood before me but averted his glance, and hung his head in shame, almost apologetically. It was evident that Jimmy feared his boss more than he claimed to love and admire him.

The man with the gun ordered me to remove my hands from my pockets. Jimmy was then led away by two henchmen, each tightly gripping an arm. I sat motionless before the boss, whereupon he offered me a cigarette. I politely declined. He didn't take off his glasses as he stared at me. He laid his pistol down on the table, blew smoke in my face, and began speaking with a thick Eastern European accent.

"My name Kosta. You have ze money of your friend."

I remained silent, not sure if this was a question or a statement of fact.

"Your partner Jimmy, he is good man. But he owe me money," he sneered.

How did I not see this coming?

"But he lucky for having comrade like you," the boss said, smiling.

With an almost imperceptible motion of his finger, one of his men came over with a bottle of Russian vodka and two sparkling crystal glasses. They made a distinctive clink. The boss poured us each a slug and downed his in a smooth movement, before pouring another one.

"You don't drink?" he asked, looking with disgust at my untouched glass.

"Of course I do," I said, downing my elixir in a feeble attempt to nullify his intimidations.

He was now sipping his third slug and took off his tinted glasses to prick me with his gaze. His eyes—I was surprised he had any—were cold, dark, and empty, framed by bags of grey skin that betrayed his fatigue. He sniffed and lifted his vodka to me.

"I like Souz Africans. Why? Because you are honest. Zis I like." His words were clipped and calculated. He was already mind-fucking me. "But you know ze problem wiz you?" he waved his finger at me.

I remained silent.

"You are too much trusting in people." He took a puff of his cigar. "See, you believe Jimmy, now you sit here wiz me. And now you vill give me money because you also trust me too much." He smiled, seemingly pleased with his deductive reasoning.

No, I'm pretty sure it's not because I trust you.

"Your name?"

"Darklay, Alex."

I saw no harm in giving my actual first name. It was easier for me to remember.

"Alex? Alexander. Good strong name. Alexander, he also a young man who conquer my country of Serbia many century ago. My name Kosta, like Konstantin, ze ruler who bring Christ to ze rest of world. You have family?"

"No, the Legion is my family," I said with a straight face, as I thought of my parents and sisters in Johannesburg.

"I not believe you. You have mother?"

"My family is dead. Car accident. Why else do you think I joined the Legion? I have nothing to lose, nothing to go back for. My father is the regiment's colonel. My girlfriend is my FAMAS assault rifle."

"You are in ze Legion of mercenaries? Courageous man," he said with surprising sincerity.

Strangely, it seemed that my status raised his level of deference. I briefly reveled in the pride of being an elite warrior, even in the eyes of a hardened drug lord.

"Alexander, you too courageous and trusting. Zis vill get you into trouble. People is born sinful. They never change. I now must give advice for you."

"Okay." Advice from a Serbian drug lord? I was genuinely intrigued.

"I am kind man. You are young. I teach you about zis vorld. You give money to anybody who say he your friend. But I tell you, you should belief nobody. You must take security for all loans. You hear me?"

He opened a small satchel lying on the table next to his pistol and held up a petite plastic bag filled with white pellets.

"Zis is like gold, wors two sousand Euros. You keep zis until you get money back from Jimmy." I took the stash from him and inspected it. There were five thumb-sized capsules inside. Their street value would be an excellent return on my investment. He watched me carefully and then continued.

"It is covered in wax so it slides easy into ze rectum. Zere are a few layers to protect ze drugs inside. So clever. It is best way to get ze good shit from here to Europe. Women are best carriers. Nobody suspect zem. When zey get zere, zey get paid ten times more for it," he said proudly, as if he was a philanthropist creating jobs for poverty-stricken women. "You give me money. You take this. We all happy," he smiled.

As I sat in that room, surrounded by men who had seen some of the darkest corners of the world, I was tempted. I could buy false papers and easily get myself home on the next ship or plane out. I knew I'd never see my money again, and wondered if this option would at least give me another hope. In a country rife with narcotics, I'd never get caught. Suriname's economy was too dependent on this trade. I'd be a fool not to engage in this Faustian pact.

I had a flashback to a moment in my teenage years. While drunkenly loitering with my friends in our neighborhood, we came across a new motorbike parked outside a home. We egged each other on, but something nagged at me personally. One of us knew how to hotwire it, and soon the group decided to steal it. "I'm out, guys," I said, walking away, ashamed, as they taunted me for being a soft cock. But it wasn't a victory for the morally superior. I didn't tell them not to steal. I walked away because I was afraid.

I cowered from the boss' overwhelming presence and slumped in my seat. Again, I just didn't have the nerve to take this offer. My voice cracked as I mustered the courage to respond in a way that wouldn't get me killed. "I respect your offer, sir, but Jimmy is a fellow South African and a comrade. He will return my cash."

His eyes narrowed, and he leered aggressively at me. As if he was playing a joke, he burst out in a guffaw. "Yes, you do have bravery! As you vish, but you vill still give me your money." I parted with my cash like a lamb to the slaughter. Jimmy returned, and we were escorted out of the room and its thick air of tension. I rejected the wishes of a man holding a gun, and expected a bullet in my back as we walked to the door. We left unharmed. "I don't need to know what you got yourself into," I said to Jimmy as he sniveled. We briskly continued down the corridor. "By the end of this month, as God is my witness..."

But before we were in the clear, I heard the distinctive clink of a round being loaded into the Zastava. Footsteps followed us down the hall. As I turned to face whatever was coming, I found myself staring right back into the same unflinching eyes of death.

"One more thing," the boss said as he approached us. Right then I so longed to be back in the loving arms of my Foreign Legion. "When I vas young man in Belgrade with talents, I was eager to rise up ranks in my syndicate. But I also very much afraid of ze big boss. One day he offer me something like promotion. He say that opportunity exist in South America. I also nervous. What if I don't do things like he vants? So I tell him no. He say to me something I never forget. 'Kosta, a smart man may know ven say no, but a smarter man knows ven to say fucking yes!'"

Rather than taking a taxi back to ZIN, I walked. One thought came to mind: *What the fuck am I doing with my life?* I was suddenly overcome with a sense of guilt and failure. I went to the Legion seeking meaning and adventure but found neither. I was drinking away years and months. I was determined not to make this period in my life akin to William Boroughs' lost years or name this chapter in my life "*Darklay Gets Wasted.*"

"You vanished again," Connor said when I found him the next morning. "Who were you with this time?"

"Just having vodka with some Serbian bloke."

The previous night's events aside, my mind was now thinking of the unthinkable.

"Dude, what are we doing here?" I eventually asked Connor.

"What do you mean? As in why God put us here on Earth?"

"No, well, yes, but I mean, what are we doing in the Legion?"

"I don't know. That's a deep question for this time of the morning."

He finished off the last of his breakfast Parbo Bier and put it down next to him on the tiles. The sound of the empty can on porcelain pierced my forehead.

"I really need an aspirin or a dozen. Got any?" I asked.

Connor shrugged. "Sorry dude, I'd recommend you drink another beer, but that's a shitty habit to start."

I took a deep breath and leaned forward, resting my forehead on my hands.

"What's going on?" he continued. "You've been in this funk since we got our jungle badge. Come on. Buck up, mate."

I raised my head and looked up at him. "Not only am I bored, but I'm slowly cracking up in the Legion."

Connor leaned back in his chair and abruptly crushed his empty beer can. "You too, huh? Shit. I thought I was the only one."

"Four more years of the damned Amazon, the mud, the assholes at the REI," I started. "We're interrogating suspected smugglers, leaving them to die, and getting completely shitfaced at night as if nothing had happened. When we're lucky, we get to guard the swamp for weeks so that France can shoot some fucking bottle rocket! Oh, and now, so that you know, I got some Serbian drug lord up my ass. This isn't what I signed up for."

"Let's start talking clearly, mate," Connor said as he was not one for beating around the bush. "What exactly are you thinking about?"

I paused.

"I just don't know…"

"Oh, I think you do," he replied, filling the vicinity with his strong presence. He must have been a shrink or a priest in a previous life, for he always maintained inexplicable composure and calmness in any situation. I couldn't bear the thought of leaving him behind if I made a run for it.

He wiped the sweat off his brow. Leaning forward, he looked me dead in the eyes: "Do you mean to tell me that you are seriously thinking about—"

"Wait, wait! It was just a..." I backpedaled, expecting to get scorned for my cowardice and stupidity. But his eyes brightened.

"Tell me, Darklay, how the hell are we going to get off this continent without our passports?"

"*We?* You mean you and me?"

He grinned.

"Well, I can't let you go out on an adventure without me! You'd get yourself killed."

My hangover suddenly cleared. For the first time since finishing at the top of my *section* in *instruction*, my life now had meaning. I was excited about something.

A legionnaire's drive for freedom was a liberated genie. As the *cafard* grew, so too did the ripening ideas of desertion. Most escapes were executed without much planning. With legionnaires being paid less than the poorest civilian, most deserters embarked by foot, for the desperation of escaping was more significant than their patience. Due to their situation, men always walked right into the hand of patrolling gendarmes. Paradoxically, the quest for adventure that drove me to join the Legion was the very thing that was driving me to leave.

It was days after this conversation that I began to run into a particularly unpleasant legionnaire. "You fucking pussy," he always whispered in my ear as he passed by. *But how did he know of my secret?* He consistently showed up when I was drinking. From across a dim, smoky bar, he stared at me with a sickening smile. "I'll buy you a drink," he finally said after gathering the courage to confront me. "You seem to be good at pushing back the cocktails these days, yeah?"

"Um, okay."

"What do you suppose everyone's going to think about you now, you fucking quitter? Do you really want to return home as the coward of Johannesburg?" he continued to harass me.

"I don't know what you're talking about. I spilled enough blood and sweat to get here, and I got the badge to prove it. I'm not about to go anywhere. So fuck off!"

He sat back down at his table, and I carried on drinking. But then he started throwing peanuts at me. Most bounced off my head while others went down my shirt. I was near the end of my tether and rage simmered inside me.

"You gave them your word," he said, instantly appearing behind my right ear. "You couldn't even complete your contract. What makes you think you'll amount to anything in life?"

I turned around to confront my tormenter but saw that now sitting at his table was Jannie himself! But how?

"Alex," Jannie called out, "you promised that you'd stay in the Legion for me…"

I stood up out of my chair, grabbed my doppelganger's throat, and squeezed.

"Why'd you bring my best mate here, and who the fuck are you?" I demanded."

"You already know who I am," he replied in a clear, calm voice, with a sick smile. "I'm Alex de Bruyn."

Suddenly I was jostled out of my trance by the sound of a shot glass being slammed on the bar next to me. Everything was normal, and Jannie and my double were gone. I swigged the slug anyway and promised myself to lay off the booze.

Escaping the Amazon

Our cash and essential belongings were still at the REI, so two days later, Connor and I crossed back into Guyane. I had a week of leave left, so our superiors were suspicious as to why I'd returned. Since the Legion held our passports under lock and key, we now had the added risk of deserting without papers. But we figured it safer to be wandering around undocumented in Suriname where the Legion had no legislative power than Guyane. There was no turning back, and so we agreed to pack and then meet outside the REI's west wall the following noon.

I woke early that morning, ready for my next adventure. My body trembled with adrenaline, but I kept calm as I waited for the legionnaires in my dorm to head out for their coffee and baguettes. Dressed in civilian clothes, I grabbed my meager belongings from my locker and shoved them into my rucksack.

At 11h45 I briskly walked out of my company quarters with my overstuffed pack on my shoulders. My heart pounded inside my rib cage, and time slowed to a crawl as if I was walking a tightrope across the Niagara Falls. I passed the officers' mess, not even glancing to see if anybody noticed me. Rookie Legionnaires were weeding the flowerbeds outside the medical center. I didn't look up; I didn't look around, and I didn't greet anyone. I just wanted to break into a mad sprint.

When I finally arrived at the gate, it towered above me, seeming more intimidating than before. I looked it up and down and reluctantly turned to the guards standing in the way of my freedom.

"Where are you going?" one asked through a cold, thick accent. We had shared a few Kronenbourgs before, but we were by no means buddies. I didn't even know his nationality, but his sad demeanor indicated that he had no choice but to be in the Legion.

"I'm leaving," I said, calmly, locking eyes with him. I was too nervous even to try to lie. He could see me shaking and sweating like a thirteen-year-old on his first date. I was expecting to be arrested and immediately jailed for so blatantly trying to desert in broad daylight. But to my absolute surprise, he smiled and opened the gate. "Well...I see. Um, good luck," he said, stepping aside and saluting me, a legionnaire with no rank.

I walked quickly to meet up with Connor. Unlike me, he was not technically on leave. He chucked his rucksack over the two-meter high wall, smoothly and gracefully pulled himself over it, and jumped down to join me on the other side.

"All those pull-ups finally came in handy," I said. "Never mind that you're taller than the damn wall!" We laughed, dizzy with excitement.

We gingerly passed by the post office, and I nervously requested to withdraw funds from my Legion account.

"You are extracting all but ten of your entire five thousand Euros balance?" the clerk asked.

"Um, holiday. I'm going to the beach with my local girlfriend and booking a nice hotel."

We had to withdraw nearly every last penny from my account because the Legion would freeze it upon any suspicion of desertion. Even a significant account withdrawal would be flagged and made known to the military police. We didn't know how quick any electronic communiqués would occur, and as such needed to leave Kourou as soon as possible.

We got into the first taxi we could find and paid the driver top-dollar to haul ass out of there. We then instructed him to take the longer route to the border. Riding in a local cab was dangerous even in the best of conditions, but doing so on the back roads and telling the cabbie to hurry was suicidal.

166

"When is *rassemblement?*" I asked Connor worriedly. "Do we have enough time before they phone the border officials? I don't think my cash withdrawal raised any flags. We'll find out soon."

I only stopped shaking when we crossed the river safely into Suriname. Only then did I give Connor a high-five.

"We did it!"

Like rebellious teenagers who manage to run away from home, reality slowly dawned upon us.

"Okay, so now what the hell do we do?"

BEN-HUR

If only it were all so simple! But the line dividing good and evil cuts through the heart of every human being. And who is willing to destroy a piece of his own heart?

Aleksandr Solzhenitsyn, *The Gulag Archipelago*

My priority was to get myself a new passport. I was no longer a legionnaire but a deserter. The closest thing I had was my Foreign Legion ID, with my *nom de guerre*. As there was no South African embassy in Suriname, my only option was to get temporary papers through the one in Trinidad, an island in the Blue Amazon six hundred miles away. Luckily Connor somehow had a second passport at home. His parents rushed it to him via FedEx Express International. Now he just needed to wait for me to get myself sorted out. The one thing Connor and I agreed upon was to somehow fly home to South Africa together. Although we were no longer on French territory, we didn't want to stay in Suriname for a day longer than we absolutely had to. France had a habit of going to great lengths at breaking international law to round up wayward legionnaires.

One noteworthy international incident known as the Casablanca Affair went all the way to the International Court in The Hague. In 1908, six deserters, three of them German, then in Morocco, were personally issued safe conduct papers by the German ambassador. They attempted to take passage upon a German vessel harbored in Casablanca. Just before boarding, the escapees were violently seized by the French authorities. The event became a major international incident between both countries. The court eventually ruled that France had jurisdiction over the deserters and was not called to surrender the legionnaires.

It was initially difficult to adjust to our new freedom. Even without *reveille*, we reflexively woke up at 06h00 sharp. We didn't know what to do with our time, and that inevitably led to problems. At University I always looked forward to the summer holiday and made plans to read dozens of books, catch up on the next term's topics, or learn a new language. By our second day in Paramaribo, we'd already stopped shaving, keeping in shape, and violated almost every point in the Legion's *code d'honneur*.

I remember watching my local news, or American escaped convict reality shows on television. I was dumbfounded at how the escapees were invariably caught only a few miles away from their maximum-security prison, usually at their girlfriend's home. I was in danger of becoming one of those dumbasses. Rather than keep a low profile in Paramaribo, we decided to rent a house instead of going back to ZIN. Word got out, and our place soon became inundated with freeloading legionnaires on holiday who wanted to party, and save money to divert it on booze and whores. Our place naturally became *The Basecamp*. With loitering local women, and occasionally, a European tourist running about, ours was a somewhat lighter version of Sodom and Gomorrah, except the neighbors kicking down our door, were drunk men from the REI. We were oblivious to the fact that at any time, any legionnaire could break a vodka bottle over our heads and drag us back to the regiment, probably for a reward.

"Do you think they'll give me any hassles?" I asked Connor one night while we were lounging on our deck chairs.

"What are you talking about?"

"The embassy in Trinidad wants me to go and sign a declaration here at the police station," I explained. "They were really friendly and sympathetic with the story I made up. I told them I was on holiday in Suriname and Guyane and lost my passport in the jungle, as simple as that. Hey, that's my story, and I'm sticking to it."

"In this place, you should be fine," he said. "If you do get arrested, I'll come and break you out."

"Hey, that shit's not funny. Unlike you, I'm the asshole without a passport."

Escaping the Amazon

The next day, as I headed out to the police station, my stomach churned with anxiety. I walked in through the wooden doors. The creaking shattered the silence, and the man behind the counter looked up from his stack of paperwork. I approached with as much confidence as I could muster.

"Good afternoon, sir," I said, thankful I was no longer a teenager with a voice that could crack at any moment.

His face was void of expression. He sat close to the counter, moving his computer mouse with one hand and eating a greasy chicken drumstick with the other. I was interrupting, and so spoke slowly and quietly.

"I am South African. I'm here on holiday. I lost my passport in the jungle. I need to send a declaration to my embassy to get a temporary replacement so that I can go home. Can you please help?"

He ignored me while he finished eating, licked his fingers, and reached for a pile of blurry forms. He took one and placed it in front of me, leaving his greasy fingerprint in the margin. An old fan in the corner of the room provided some relief from the heat, but the smell of stale smoke had penetrated every exposed surface.

"*Schrijven*," he said, tossing me a broken BIC pen. I obliged, indicating that I had lost my passport—detailing exactly how it "didn't quite" happen. Without reading my account, he stamped and signed the sheet. Lifting his large body from the small wooden chair, which creaked with momentary relief, he made a photocopy and handed me the original.

"Thank you," I smiled, and calmly walked back out onto the streets.

Shit, that was easy! I thought to myself as a wave of relief swept through me. I happily made my way back to *The Basecamp* where Connor was actually surprised to see me.

"I wasn't joking when I said I'd come to break you out," he muttered.

I immediately sent my declaration to Trinidad and commenced the game that one always plays with South Africa's glacial Home Affairs Department. In my best Tom Petty, voice I sang: "*You take it on faith, you take it to the heart. The waiting is the hardest part…*"

Connor and I spent those days figuring out what we'd do next.

Despite our partying and minor celebrity status, the days began to blend into one long continuum of boredom spiked with debauchery.

"The devil's play tools," Connor joked as he mixed himself a cocktail for breakfast.

"Put that down. I want you to understand this. Just because we're out of the Legion, it doesn't mean that I'm no longer craving adventure, utility, or purpose—anything to fill this void."

Among the stream of semi-recognizable legionnaires wandering in and out of *The Basecamp*, one stood out. Vasile was one of the few who thanked us for letting him crash at our place. He was slightly shorter than me, with a dark complexion, and was one of a handful that spoke English and could seamlessly drift between both *mafias*. Vasile was a fit and squared-away legionnaire, embodying the most excellent attributes of our corps. He also turned out to be an unbelievably loyal *camarade*. When he overheard us discussing our plans, without hesitation, he petitioned to be part of our folly. He wouldn't have deserted with anybody but Connor and me. Vasile was a reincarnation of Jannie. He was a follower and a pleaser, and we were happy to take him under our wings. In a touching gesture, he confessed that he had no other mates in the Legion but us. "You're all I have…"

Although the introverted Vasile was not someone I would have befriended as a civilian, in the Legion, one can't be picky. The poor lad was completely awkward and uncomfortable around women, and only mustered the courage to approach them when he was drunk. Prostitutes were the exception, and he was an addict and sucker for them. When paying for sex, he was all alpha. "My beautiful girl has so much going for her," he had told us one evening when they came around for pizza. "I think I'm falling in love with her."

"She's definitely head over heels for you," Connor joked, reaching for a slice.

"I need to rescue her. She's working against her will as a sex slave. Men are behind this trade. I'll be her protector, provider, and a one-woman guy, you know?"

We did but also knew the inevitable. This may have been his first rodeo, but it wasn't hers. Three days after he stopped paying her, and the sex was "out of love," he became a whimpering, blubbering teenager who couldn't string two words together in front of his "eye-rolling" girlfriend. Inevitably, the relationship ended, and Vasile went back to spending his money on satisfying his urges—and finding a new damsel in distress.

However, Vasile was waiting to receive his special pay-bonus from the Legion at the end of January, and so had to stay there for another month, while Connor and I got our plans together. Like me, Vasile didn't have a spare passport. As such, we agreed to touch base again once he was ready to join us. With his holiday at an end, he returned to the REI.

By phone, it was impossible for me to get hold of a real person at the embassy. The days quickly turned into weeks. Letters, emails, and faxes went unanswered, and it was eventually evident that my passport was not going to materialize. These problems were *de rigueur* for most Africans abroad.

Escaping the Amazon

Connor and I needed to come up with Plan B. Patience wasn't a virtue we beheld. Haste makes waste, but we simply couldn't carry on in Paramaribo any longer. We needed to get off the South American continent before word got out to the Legion about the rock star deserters thumbing their noses towards the REI just across the border.

"Dude, I can't board any plane without a passport," I pleaded with Connor.

"Okay, well, what if we just make a raft and row out of here? We did that in our Jungle Course."

"That's the stupidest shit I've ever heard. Really?"

"How about we just buy a jalopy and drive to Venezuela and get you to the Embassy there?" Connor responded with what seemed like a better idea.

"Hey, Einstein, how will we cross that border without my passport?"

"Relax, fine, then why don't we instead procure a sailboat and cruise out of here? It's a little smarter than the 'dumbass rowing' idea," Connor proclaimed as he sat up straight. "I've done a fair bit of sailing in open waters. I'm pretty sure we could make it."

"'Pretty sure? Where'd you learn to sail?" I asked.

"I use to with my dad. Sails are down. Sails go up, and the boat goes forward. It's that simple," Connor explained gesticulating with his hands. "When we hit land on whatever island, we just say that the bag with your passport was washed overboard and that we simply need to get to the embassy for a replacement. If they deny us entry, well just drift to the next island."

The economy of Suriname was still based mainly upon bartering and private sales. Manufactured and assembled goods had to be purchased abroad. There was little public information to help us buy a sailboat. Days of sweaty footslogging and bickering with local fishermen were fruitless.

"Another one of your great ideas, Connor. How about you just fly to Trinidad, buy a boat, and sail it back down here. You can then show me how easy it is to drive it, you know, sails down, sails go up, the boat goes forward," I ribbed him, as I was still trying to piece together how much experience he really had.

"Brilliant idea, Darklay."

"You serious?"

And so we had a new plan. After Connor flew out to Trinidad, I phoned the embassy again, just in case. They'd forgotten to tell me to get fingerprints taken, and instructed me to go back to the police station to do so. I hesitated, but after my first smooth encounter at the police station, I wasn't worried.

By now, Connor and I had paid a month's rent in advance on *The Basecamp*, still teeming with Legionnaires. Although we trusted each other with our lives, legionnaires didn't trust anybody with their money, and so I stuffed all my cash and a cell phone into my rucksack before heading to the station.

Arriving there, I walked inside. The same greasy-fingered official sat behind the counter engrossed in coffee-stained paperwork. I asked where I could get fingerprints taken and he nodded towards the staircase on his left. I went up and saw a sign that read FORENSISCHE AFDELING—FORENSIC DEPARTMENT. I poked my head into the small room.

The officer inside turned around. His uniform was perfectly tailored, immaculately ironed and tucked in. He was short and stocky with a round brown face and a thick head of cropped black hair. I noticed a shiny name badge pinned to his shirt: Scholten. Without any expression, he spoke in broken English.

"Why need you fingerprints?"

"I am South African. I'm here on holiday. I lost my passport in the jungle. My embassy needs them." He didn't motion me into the room.

"Show me your papers," he said in a mistrusting tone.

In a folder, I had all my documents, including a copy of my South African identity certificate. I handed it to him, but in doing so, something tumbled out. My heart stopped and time froze as I watched my Legion ID land face up on the floor.

Shit, I thought, starting to panic.

I nonchalantly bent down to pick it up, but the officer was quicker than I and snatched it before I could. *I should have left this behind when I quit the Legion. Why the hell did I even bring it along?* But he was not even looking at it as he held it out for me to retrieve.

I reached out, trembling. Then, he stopped and brought it up to his face to inspect.

Please, no.

At first, he was confused, then his eyes widened as he again checked my South African identity document.

"Who this?" he said pointing to my Legion ID. "DARKLAY, Albert" was clearly printed next to a picture of me in uniform. I should never have accepted my *nom de guerre*. For some, the new identity saved their lives. For others, it was a cool James Bond perk. For me at that moment, it was about to get my ass locked up.

"That's me," I said.

"And who *this*?" he asked, raising his voice as he pointed to my South African identity document.

"Well," I had no defense, "that's also me."

In a split second, Scholten, who was half my size, grabbed his handcuffs from his top drawer, whipped my hands behind my back, and cuffed them. He didn't want to risk wrestling me. He swiftly ordered me to sit down and telephoned his colleague down the hall. Back-up promptly arrived, and Scholten rattled off a list of orders too quickly for me catch They each grasped one of my arms, pulled me up from my chair, and escorted me down the corridor. I was placed in a bare cubicle with a single wooden table and a chair. The floorboards groaned as we walked inside. No windows—I knew this was for interrogation purposes.

Scholten ordered me to sit before the table and read from a notebook in English: "You have no passport. You are being arrested. You have the option to request that we alert your embassy. You have six hours to be questioned and to state your case."

His assistant removed my handcuffs, ordering me to place my hands in front of me. The table had a thick metal handle bolted into the wood. He looped the cuffs through it and re-cuffed me tightly. They both left without a word, leaving me chained to the table.

Five hours passed, the cold metal handcuffs turning my hands blue. The empty room reeked of armpits. The wooden table was stained with coffee and dried blood—it silently taunted me with the magnitude of what had just happened. Sweat dripped down my forehead, burning my eyes. My body ached from sitting hunched forward for so long, but my soul ached even more. My mind spun with ideas of how to break out of this joint.

All I'd been able to hear for the past hours were the faint Dutch voices from the station entrance downstairs. I looked around at the peeling paint and reflected for a moment on the night in Marseilles that had brought me to this point. I was in search of a dream—a dream that now seemed shattered. The walls moved in closer and conspired with my flashbacks. I would pay a hefty price for an anticlimactic and misdirected hunt for purpose—not to mention my alcohol-induced absentmindedness.

I then heard purposeful footsteps coming down the passage towards me. They stopped at the door, and Scholten entered. He was preceded by a sharp nauseating smell of cheap cologne, spiked with that of unwashed feet, strong enough that one wouldn't dare leave food lying around in that space. But by now, the stench may actually have been my own. Accompanying him was an interpreter. The chairs squeaked on the wooden floor as they pulled them out and sat opposite me. Scholten was even more gruff and fidgety.

"*Wat is uw naam?*" he spat out in Dutch. His voice was as abrasive as sandpaper. Too many cigarettes, it seemed.

"What is your name," came the translation.

"Alex de Bruyn," I answered, not looking at either of them. My eyes were on my cuffed hands.

"*Wat is uw nationaliteit?*"

"What is your nationality?"

"South African," I said.

"*Waarom heb je een Frans indetiteitsbewijs?*"

"Why do you have a French identity document?"

"I am a member of the Foreign Legion in French Guiana."

Technically this was a lie, as I had deserted. I was no longer under their protection but had no option but to stick to my backup, half-truth story. I knew enough of Surinamese justice to suspect that I would neither be given a phone call, lawyer, or a fair hearing. The said system was not based on the rule of law—my first experience of crossing the border illegally had taught me this.

They relentlessly questioned me, but no matter how hard they tried, I stuck doggedly to my narrative. It should have been simple enough—all I needed that morning were fingerprints. And yet, I was still struggling to understand what exactly they suspected me of. *Did they have an agreement with the Legion to arrest deserters? Did they suspect I was a forger?*

At some point in the questioning, Scholten and his assistant whispered amongst themselves in Dutch.

"*Hij is een Frans Militair. Hij werkt in Suriname.*"

"No! I'm not working in Suriname!" I protested, overhearing the conversation, frustrated that they had my story wrong.

"I am on holiday in Suriname, and I lost my passport."

"*Wat zegt hij?*" Scholten asked his translator, although I knew he understood some English.

"*Hij is niet hier voor werk—en zijn paspoort was gestolen.*"

Again, the wrong information: "My passport was lost, not stolen. It's probably sitting with a bunch of assholes across the border," I lied, exasperated.

Scholten suddenly jumped up and slammed both hands on the table. "*Hoeveel talen spreekt je? Je bent een spion. Welk gedeelte van de overheid bent u hier voor om over te nemen!*"

His face was red, his temples pulsing, and spittle flew in every direction as he fumed. But he was also excited at the thought of catching a French spy who was in Suriname trying to overthrow the government. I wasn't yet aware of the seriousness of this charge, and it didn't help my situation when a burst of laughter escaped my lips.

"Why you here?" he said in English with a thick Dutch accent, shoving his stubby finger at my nose. "Who sent you?"

"I told you. I go by a *nom de guerre* while on duty. But as of now, I'm here on holiday under color of my South African passport." I somehow managed to remain calm in the face of his rage. My instinct was to rugby tackle him and rain down some fist-bombs.

"You lying!" he yelled, glaring at me through narrowed eyelids.

"*Waar ben je de grens overgestoken?*" he asked.

"Where did you the border crossed?" the translator continued, aping the stubbornness of his officer, who was leaning over the table on both hands, staring me down.

"Albina," I said, remembering how easy it was to cross the frontier amidst the chaos of the river, thankful that there was no way they could track my mode or date of entry.

"Did you have your passport stamped?" asked the translator before being prompted to do so.

"Yes," I said, a measured lie, hoping they would never know any different.

"Ah, good!" said Scholten smugly, sitting down again. "We can check the database."

The database? I thought with disbelief. The entire Surinamese government was void of even a solar calculator.

I soon found myself in the back seat of the blue and white police car with my hands re-cuffed before me. My two escorts sat in the front, so I was able to slide my phone out of my pocket and send a message to the guys back at *The Basecamp*. They too had crossed into Suriname illegally. If the cops took me to the house, everyone would be in the shit. I hurriedly texted them: "OMG I ARRESTED. GET FUK OUT OF HOUSE!!!" I managed to send it just before we pulled up to the customs office.

Scholten's assistant opened the door and walked me inside. There were two desks, each with a shiny computer on them.

"We need to verify this man's entry into at Albina," he ordered the government clerk

"Certainly." She smiled at him and sat down. She then looked at me with an instantly dour, demeanor.

"Name."

"Alex."

I instantly received a slap across the back of my head.

"Full name, idiot," Scholten's assistant growled.

Even in handcuffs, I was a hairsbreadth from head-butting him and crushing his balls with my knees. While the Legion could slap me all day, I wasn't about to let a ten-a-penny bureaucrat do so. Nobody else kicks my dog but me. But this wasn't the time.

"Alexander Angus de Bruyn," I mumbled.

"A-L-E-X-A-N-D-E-R," she pronounced each letter as she punched it into the keyboard.

I knew as well as Scholten that this clerk wouldn't find my name in the system.

"No," she declared, pursing her lips, shaking her head. "You are not here."

"But I had my passport stamped," I said in feigned protest.

"Our modern system is ultra-high-tech," she said, sitting upright, offended by my cheek. "Computers never make mistakes. You did not go through the customs office at the border."

With even the low-level clerk calling me a liar, I knew things wouldn't turn out well.

"I don't doubt the system, but perhaps the official forgot to enter me into the database." I was naïve enough to think that a white South African might just be above their rudimentary legal system.

Back in the car, Scholten, sitting in the driver's seat, looked at me through the rear-view mirror. "Why are you really here?" he asked again, but I remained silent. Blood rushed to Scholten's face, spinning him into another rage. Racing back to the station, he jerked the police car around a corner. Unable to use my hands for support, I was thrown against the back door. "You try to run, and I kill you!" he shouted. I maintained my silence and reached again for the phone in my pocket. To our faithful Anna, I texted: "I BN ARRESTED. AT POLICE STATION. PLS COME NOW!!" With Connor out of the country, she was my only hope. Scholten finally caught me using my phone as we rolled through gates to the police station. He slammed on breaks, and I flew forward, smashing my head against the front seat. My phone fell to the floor. He got out, opened my door, and dragged me out of the car, tearing my shirt in the process. I sheepishly let him march me back to the interrogation room, knowing that he now had the upper hand.

Emboldened by this evidence, Scholten shoved me back onto the same wooden chair from earlier and scrolled through my phone texts.

"Are you dealing in drugs? Who are you working for? You will not get away with it!" Scholten now shouted in Dutch. I tried to speak while thinking of a way to break out, but my mind couldn't think straight. Just then, a young man came into the room and whispered into Scholten's ear. Scholten told him to wait with me and left the room. I could hear Anna's voice down the corridor. *That was damn quick*, I thought. She was the only person on the continent who could vouch for me, and what a sight for sore eyes!

"Please officer. This young man is my friend. He has done nothing wrong. He is here with me on holiday. Everything he says is true."

"Why doesn't he have a passport?" Scholten asked, sounding pacified by her presence. Anna's motherly charm never ceased to amaze me.

"He left it in with the Legion. He travels on his military ID."

Shit. We had different stories. Anna was trying her best but didn't know what I had already told them and was jacking it all up.

Scholten's calmed state evaporated quickly. "You are both lying! You think I am stupid?"

He ordered her to leave, but she persisted, telling him that she would assume responsibility and immediately return me to the Legion.

"No. He goes nowhere."

She was escorted out by an armed guard, and I didn't even get a chance to see her robust but majestic presence.

"It is Friday. You can only go to court on Monday. You will go to the cells," he spat at me.

The cells.

What I thought would eventually be an amusing yarn at a Johannesburg dinner party was no longer funny. My head dropped with the weight of images of *Papillion*, *The Dry Guillotine*, and *Midnight Express*. I froze with fear as I watched Scholten empty out my bag, count my cash, and confiscate my electronics.

"Welcome to Suriname, your playground, European bastard..."

PHILIPPI

I was always astonished at the extraordinary good nature and lack of malice with which men who had been flogged spoke of their beatings and of those who had inflicted them.

Fyodor Dostoevsky, *The House of the Dead*

It was Friday night, and I was forsaken. *They're locking me up. They're locking me up. They're locking me up.* I repeated those words in my mind, willing them not to be true. A warden escorted me down to the cells on the north side of the police station. We walked down a long passage to a white steel door that opened towards us. Past that was a steep flight of narrow concrete stairs. The cells were underground and hidden from the world. The smell of ass and mold assaulted me as we descended into the first level of Dante's *Inferno*. All I could see through the damp darkness was a single, low-hanging light bulb. As I got closer, I saw a warden sitting directly under it at a small desk. I locked eyes with him, but he simply stared through me. To my right, I could only hear movement from the living beings within.

I signed a register in silence. The officer then ordered me to remove all my clothing. Thinking I had misunderstood him, I stood idle.

"*Verwijder al uw kleding!*" he repeated in a shout.

Escaping the Amazon

I had little choice in this matter. With all my clothes lying next to me on the ground, I stood naked for an agonizing few moments while the warden squeezed his right hand into a latex glove. The distinctive "snap" sound was the stuff of nightmares for any red-blooded male. I was told to squat deeply so that he could search my most intimate areas for drugs. It took all I had to restrain myself from beating him to a pulp, not only for violating my body but for having fingers thicker than those of a baseball glove. Once the investigation was over, I dressed again, still aching from both the invasive digits and the humiliation.

"Don't make problems with your new friends," he said, fumbling with a bundle of archaic keys.

As I walked further with my empty bag in hand, a massive gate behind me slammed shut. The clank resonated with my imperfect soul and sounded like that which the damned heard after final judgment.

Slowly, I strode into a sea of bodies. The prisoners gathered to glimpse who had arrived, as I pushed through the crowd. I scanned the multitude of scarred, bruised, and sunken faces and couldn't fathom what had brought them to this Black Hole of Calcutta, or what thoughts were running through their minds as they sized me up. I was outnumbered, and the silence was deafening.

Suddenly, cold fingers gripped my right arm, and my ribs seared with pain from a blow that came out of nowhere. My reflexes were quicker than my mind, and in a split second, I had grabbed my assailant's neck and smashed him up against the wall, ready to land a crushing elbow to the head. Everything seemed to happen in slow motion. My adrenalin surged, causing ringing in my ears, but this was all interrupted by a burst of laughter from my assailant. I was twice his size and stronger, but there was no trace of fear in his eyes. His response disarmed me, and I let go of him, trying to figure out what the hell was going on. He was scrawnier than every man standing around us, with a teardrop tattoo under his eye. His swagger signaled that he was in charge of this joint, and his well-timed guffaw perhaps even saved my life. He slapped me on the shoulder and announced: "You're bunking with me."

The crowd dispersed quickly at his command. I was too confused even to argue, and so followed him into the first cell on the left with no windows or natural light. In one corner there was a shrine to the television gods—a stack of DVD's taller than me—a hammock, and bunk bed squeezed into the remaining space. He walked across the cell to the hammock and tipped it over, sending his sleeping cellmate tumbling to the floor. I was announced as the new occupant, and the previous tenant silently scurried away.

I was still on guard against rape, a beating, knifing, or worse. But to my surprise, the boss seemed only to want to talk. I hesitantly sat down on the hammock, and he followed suit on the bottom bunk. The top bunk was disintegrating and hung over him like a collapsing tent, but he didn't mind. Our exchange was slow as we strung together phrases from the few words we mutually understood.

"What is your name?" he asked me in Taki Taki, the local dialect, some of which I had already absorbed by merely being in Suriname. The only soft skill that the Legion taught me that might get me into a Top 20 Business School was the ability to learn new languages on the fly.

Leaning forward, the boss pointed to his arm, which was draped in a blanket of tattoos from his shoulder down to his wrist. Two words stood out boldly amidst the skulls, hearts, and faces: "Al Pacino."

"They call me Al Pac."

He lay back onto his mattress, lacing his fingers across his abdomen, bloated from malnourishment. "Why are you here?" he asked, with a piercing stare.

"Man, just passport issues, but I'm going to court on Monday," I said, trying to convince myself this was true. I vaguely estimated that he was somewhere in his forties. He was bald, and his inked skin hung loosely on his wiry arms and torso. His eyes were dull and emotionless. This was the house of the dead. I took a chance and returned the question to him.

"I'm here because I didn't kill two people," he said with a shrug.

"So you and everyone else in here are truly innocent?" I asked sarcastically, taken aback that he had actually shared anything with me.

"No. There were six who deserved to die. I killed four, but when I got to the last two, I felt remorse and let them go. They went to the police. You see?"

I didn't know what to say and just lay back on the swinging hammock, allowing its rhythm to numb my mind. Al Pac seemed to have fallen asleep. I lay awake, wondering if the lights were ever extinguished. In Solzhenitsyn's memoirs, even during food shortages in the empire, not a single light in the gulag was ever switched off. Like a drug addict, this prison seemed to live a 24-hour cycle. My new friend Al Pac slept like a baby.

I lay there for several depressing hours remembering the French countryside in the spring and imagined Jannie grilling a steak with me under a tree. "You shouldn't have gone to France," he pleaded with me. And then he returned a perplexing worried look. "Alex, somebody's coming…" I was suddenly awoken from my light slumber by the warden's brisk footsteps and the echo of our cell being unlocked.

I sat up, dazed, my mind still trying to make sense of where I was. I then became aware of the concrete floors, the hammock, the bunk bed, and the light in the passage still humming imperceptibly. The sound of radios, televisions, conversations, porn, and emaciated bodies rolling in creaky bunks intensified. I even perceived a slow drip from the ceiling across our cell. It was as loud as an airliner at takeoff.

There was a sudden bustle of movement behind me as Al Pac jumped out of bed. He whipped on his clothes and marched with a plank of wood down the corridor. "Shower time! Shower time! Wash your asses!" he shouted with excessive zeal, as he hit each bar of the cells, causing a sharp clang.

Slowly, the cells came to life, and I saw shadows standing upright. The others—dull, lethargic, and depressed—displayed none of Al's energy and enthusiasm. He continued to shout and bang until every man was accounted for, even those still high on drugs or glue. Like cattle, he herded them all to the showers.

Al Pac seemed as effective with his type of men as any Legion NCO. He had his own private army and was the self-appointed *subcommandante*. I joined the herd towards the showers, not in obeisance, but because I'd been holding my bowels out of fear. With that fright gone, I needed a toilet fast. I continued down a long corridor, lined with unlocked cells. The air smelled of balls, marijuana, and mold. Considering my invasive frisk the day before, I couldn't imagine how any drugs got in. I then wondered why prisoners were all hesitant to wash but only momentarily.

The washroom was an open plan shower and toilet area. I vomited a bit into my mouth as I jumped over puddles of runny shit to get to the pots on the right—two toilets for three hundred prisoners! I squatted over the seat, caked in decades of biological matter undiscovered by modern science. Out of the wall opposite me, a constant stream of water poured onto the floor. That was our shower, and it flooded the entire vicinity with human excrement. I sloshed through it towards the small basin, regretting not first taking off my shoes. I forced myself to drink from the rusty tap, trying not to get my mouth too close to the stagnant grey water stopped by a hair-clogged drain.

After our morning ablutions, we were given breakfast in our cells: one piece of stale bread with a dab of chili smeared onto it. Apparently the prison kitchen—if one could call it that—believed that chili was sufficient nutrition to keep us alive. My eyes watered from the spice, but I tried to savor the experience of chewing something rather than gulping it down. Surprisingly, they followed with a cup of tea. It was ironic that prisoners who had to wade through shit would be offered such a treat. If we looked hard enough, there might even have been a chocolate-mint under our pillows. With or without milk, my hot cup of English breakfast tea brought me back to sanity, to the breakfast table where my impeccably neat father would be waiting for us with fried eggs, bacon, and leftover Pap and Boerewors warmed with a tasty tomato relish. I finished the last drop of happiness, not knowing that it would be our only sustenance until the next morning.

I made my way to a common area where a soccer ball was being aimlessly kicked around. Some prisoners were chatting with each other, but most were just sitting, waiting, or staring into nothingness. The ghost was no longer in their machine. I was told that God would never give us more than we can handle, but was that really true? I hesitated to question whether there were situations where death was preferable to life. I had no idea what lives these men had lived, and with what permanent emotional injuries they were walking. There may have been equally blank apathetic faces in Johannesburg gazing at a television or computer screen. By hook or by crook, death and anguish made me more lustful with hope. I wanted to bring them the good news of life, but I was doubtful myself. In due course, I could have ended up wallowing in despair like them. Even Solzhenitsyn put in a good day's work—for the first week. Me today; you tomorrow.

Escaping the Amazon

An old man shuffled past me, looking like he had been mistakenly placed in prison rather than a retirement home. He was the only one who'd been there longer than Al Pac. No one knew why he had been locked up, or for how long. As he gingerly sat down, I could hear his joints groan under the strain of a hard life. Though he was a grandfatherly figure, I doubt he'd been so tender as a young criminal. What a disgrace to spend one's final living years in this undignified pit. I felt like I also deserved to be sitting in this self-inflicted purgatory. I was a sinner and had brought this dishonor upon myself. I had given France my word and then ran off when the going got too tough. In Dante's Inferno, the innermost circle of hell was reserved for traitors. I also began to lose my sanity. My self-doubt made me want to weep, scream, and vomit. And like the grand traitor Judas, I too was tempted to end it all. I wondered if he had repented in those immediate moments after he jumped and before the noose tightened. Would God mitigate my fault under the grounds of insanity? Hell truly was a place behind a door that was locked from the inside.

With no natural light, time was a relative concept. Curious prisoners came to chat, but I was careful not to reveal too much about myself. Al Pac buried his soul in marijuana. He admitted that he'd breathe the smoke like oxygen if he could. Almost everyone around me was continuously high–for the same reason legionnaires are always high: desperation and depression. Weed was the only substance that made Al Pac feel alive, he claimed. If he was awake, then he was high, and if he was high, then he was obviously alive—jailhouse existentialism.

"I tried Christianity," he said gesturing towards a crucifix hanging in the corner of the cell. "Deep in prayer once, I felt God's amazing presence. But now I can't feel His presence without a simple joint."

When Saturday evening finally came, Al Pac and I were locked in our cell again, yet all the other cell doors remained open.

"Why do they lock you in every night?" I asked Al Pac, who was sitting on the floor taking a last toke from his joint, paging through a wrinkled porn magazine.

"I am classified as dangerous," he said blowing out a cloud of blue smoke.

"Then why'd they allow me in here with you?" I followed up.

"Because you're also classified as dangerous," he said matter-of-factly. "You're military-trained."

My cover was blown, as was my legionnaire status. Now I had to, at least, use this to my advantage. Al Pac wasn't threatened by my experience, though. He'd probably dismembered and eaten tougher men than me. Yet he was fascinated by the Legion and kept me up late into the night asking what it was like.

"Tell me about basic training. I want to be a legionnaire too. Can you teach me?"

He jumped up from his bunk and pulled out an old army picture book that he kept under his bed. His thoughts about the military were as romanticized as mine had been before I joined. He lit another joint, lay back on his bunk, and I poured out my reflections on the precipitous months preceding my desertion. It was refreshing for a sinner to talk heart-to-heart with an even bigger sinner.

"And after Myshkin was abused like that," I continued, "I knew I was serving with some sick mother fuckers." Yet Al Pac was un-phased as if sodomy with a broomstick was the ordinary course of action for snitches or random victims. Though he had listened patiently, he was only mildly moved by Jannie's death. Perhaps it was the language barrier, or just maybe, Al Pac would never be able to relate to my world. A farmer's son and childhood friend of mine had once adopted an orphaned wolf. Very soon after raising the pup he came to the sad conclusion that this noble creature would never, ever be domesticated.

"You could still become a military man, one day," I naively said to him. He then explained why he was so fascinated by it all.

"You don't understand. When they move me from here, I'll be placed in a maximum security prison, where I'll never see the light of day. Unlike you, who may eventually write about your anguish, my thoughts and pain will never leave these walls..."

I began to nod off when Al Pac muttered a few final words.

"Alex," he whispered, "when you get out of here, are you going home?"

"Yes," I said, hoping it would be soon.

"If I ever break out of here, I'll visit you in South Africa. And you will organize me some guns so I can rob people. Yes?"

"Deal. Your wish is my command, Al Pac," I chuckled.

Sunday morning came, and I found myself sitting on a small bench outside my cell, eating my bread with chili. A young Brazilian approached and spoke to me in Portuguese. Like all the men in there, he was emaciated and had sunken eyes. He stood with drooping shoulders and talked in muffled sounds, gazing only at his feet.

Another inmate from the opposite cell who was lying on his bunk gingerly stumbled over to translate Portuguese into Dutch for me. The Brazilian had been inside for two years for not having a passport and hadn't even been given an initial court appearance. I couldn't believe what I was hearing.

"He wants you to break him out of here. He knows that Special Forces men like you are trained for this type of thing."

"Not so fast, guys," I said putting out my open palms. "I'm not supposed to be here at all, and as soon as they figure out their mistake, I'm gone."

Seeing the disappointed look on the young Brazilians face made me feel cowardly, like the time my armed and well-fed *section* left the smugglers to their own fate. I actually had no idea when or if I'd ever be freed, but I felt incompetent and was wary of showing it. Though I was a highly trained soldier and an expert at ironing creases in shirts, I hadn't smuggled any C4 explosives in my ass, and there was only so much I could have done.

My Legion spirit eventually kicked in, and I forced myself to unplug the pain and displace the despair with small goals. For hours, I sprinted up and down the passage, stopping to do pushups and pull-ups. I'd then mix it up with a few burpees here and there. I threw myself into it, expiating negativity from my soul like the sweat from my pores. "Goodbye old Europe, may the devil take you. We need sunlight and space to rebuild our bodies," I proudly sang out loud. A few waifish teenagers who had never eaten a full meal in their lives approached me and asked how they too could build their bodies.

The teens later spoke glowingly to the other inmates of my ability to whip them into shape, and before I knew it, I had a small group of modern-day disciples. One man even brought a notebook to scribble down my every word furiously. An overweight fellow then approached and begged me to make him skinny. He then went back to his cell and, in the midday tropical heat, returned dressed in a full-length fleece tracksuit.

"You trying to kill yourself?"

"But you say sweating is good," he answered, patting his belly. "It will melt my fat."

"But, technically sweat is made of water and…oh, fuck it. Keep the tracksuit," I shrugged, and we began with wind sprints up and down the corridor. Before long, we had even more followers, everyone sporting the warmest tracksuit they could find.

Some guys found some scrap metal bars to which they tied bottles of water, and within a few hours, the prison was transformed into a gym. No one knows that the CrossFit movement was invented by a South African in the French Foreign Legion while in a Surinamese jail. The human spirit, even amongst the worst of us, was truly a thing to behold. To think that I had ever contemplated ending my life. I could almost hear Louis Armstrong singing "What a Wonderful World."

Jail wasn't complicated. One ate; one slept, and one showered. If we obeyed commands, then we had no problems—not unlike the Legion. In this poke, Al Pac was the seasoned *caporal-chef*. One rule was that he ate while everyone else showered, and vice versa. I wondered why no one questioned this, but it wasn't long before I got my answer.

It was Monday morning, and after Al Pac's usual *reveille*, a tough-acting *jeune* was brought in, just before shower time. In marking his territory and establishing himself above the rules, he chose to sit and eat with Al Pac while the rest of us showered. I watched from the queue to see what Al Pac would do. Nobody ever ate with Al Pac.

He politely informed the novice that it was time to shower. Instead, the kid deliberately turned his back to light a cigarette. "I will shower when I want to shower."

I watched as Al Pac calmly walked back into his cell and returned with his large plank of wood. Before the *jeune* saw him coming, Al Pac slammed it into the side of his face, knocking him to the ground. Completely composed and in control, Al Pac took the plank back to his cell and repeated: "Shower time." Gushing with blood, the novice hurriedly got up and joined me in the queue. Nobody flinched or said a word, and that was that.

After breakfast, while Al Pac was showering, the wardens called me to the cell gate. "You go to court today." Relief flooded every bone in my body. *Anna must have worked her magic!* I finally saw natural light and my eyes squinted. Hand-cuffed and escorted by three armed guards, I was thrown into a vehicle.

All I could think about was how to escape. I considered making a run for it, or somehow using my cuffed hands to throttle the guards. But I couldn't bring myself to cut the servant's right ear and instead relied on what I'd always seen as the flimsiest of notions: faith.

We pulled up outside the courthouse, a rectangular three-story redbrick building topped with a black and white mansard-style roof. The large windows were covered with dark wooden shutters. The guards escorted me up a short staircase to the entrance. Still handcuffed, I sat in the corridor on a wooden bench watching people come and go. An hour later, I was called in to appear before the judge.

I walked into the musky courtroom and stopped at a demarcated circle etched in chalk on the crusty carpet. The judge sat behind a broad table, with a disinterested bailiff dressed in a police uniform at his side.

"Name?" the judge asked in Dutch, staring at me through his thick spectacles.

I answered in English. I was sure that a man with his education would understand me.

"Do you have a lawyer?"

"No. I don't feel that I need one for a lost passport."

"As you wish." He looked down and lifted a sheet of paper.

"So, you have two identity cards with different names. You do not have a passport because you claim that you misplaced it. There is no record of you entering with one. This means that you entered Suriname illegally and are here to perform subversive acts."

I knew better than to argue with the judge and assumed that it would all be cleared up in a subsequent hearing. He lowered the chargeing document, removed his spectacles and shrugged. "We will reconvene in thirty days." He slapped the sheet onto the table with two flat hands as if to say case closed.

He was about to get up and leave. Recovering from my shock, I called out.

"Your honor, yes, I do need a lawyer."

"No. You gave up that right."

"Sir," I retorted, unwilling to give up that easily, "as you surely know, in thirty days I still won't have a passport."

Suddenly, in a fight or flee mode, I reached for any legal or moral argument to persuade him.

"Your court will eventually have to deport me. They need not go through all that cost and trouble. I'll voluntarily deport myself and pay all necessary and special fines."

He was taken aback by my boldness.

"Our judicial system does not work on fees," he said.

Looking him in the eye, with more confidence than before, I replied calmly and deliberately: "Your honor, in your deep understanding of the law, I'm sure there is a fine that fits my peculiar circumstance. I'd be happy to pay it."

His upper lip twitched, and his face went from pink to red.

"Are you attempting to bribe me?" his words come out, slowly.

We were playing high-stakes poker and neither wanted to show his cards.

"In *no way* am I trying to bribe you. I respect the sound laws of your country," I responded, without skipping a beat. "I'm simply working with you to facilitate, or expedite, this process in the overriding interests of justice." What I thought were worthless years of grammar at school, were finally paying off.

"I will get back to you," was all he could muster, apparently suddenly bored with poker, and he gestured to the bailiff.

When I was delivered back to my cell that evening, still bewildered by my long day, I wasn't expecting the onslaught I received from Al Pac.

Escaping the Amazon

"Fuck you, Alex!" he shouted as he rushed up to my face the moment the gate slammed behind me. "I bring you in, treat you like a friend, take you under my wing, look after you like a brother, and you didn't even fucking respect me enough to say goodbye!" With every accusation, he launched himself at me as if to punch me. His eyes were fierce with rage.

I should merely have apologized, but after being jerked around by the justice system, I responded to wrath with wrath. With two meaty paws, I shoved him hard into the wall. "Back the fuck up," I said, offensively towering over him, "I will rip your fucking head off!" But Al Pac wasn't the problem. He was just the spark, and I was the powder keg. I just wanted to smash someone's face in. Anyone's would do.

"You won't dare touch me!" he shouted but in a position of weakness, starting to calm down. "Because then you will be sitting here for the rest of your life for *my* sentence. Am I worth a black teardrop to you? Go ahead..."

My heart raced with adrenaline, and my knuckles were white. I stared at him, finally realizing why he was the way he was. Unlike me, he had absolutely nothing to lose. To him, death would be a charitable release.

"Now I realize why the others won't touch you. Look, I know you want to die, but I'm not going to let that happen on my watch, and I sure as hell won't let you use me as your means of suicide. Shit, I don't even know what I'm saying..." I fretted looking up in despair, "What I *do* know is that there's more out there for both of us. Let's just hold on a little longer."

"You can sleep on the floor now," he said as he tipped over my hammock while the guards locked us in. I didn't argue and threw my bag in the corner as a pillow to sleep on later.

Al Pac lit his evening joint, lay on his bunk, and stared at the ceiling.

"You know," I said, "when you can no longer put a joint in your mouth, you may end up using a loaded gun instead."

"Did you ever get into difficult situations in the Legion that you regret?" he asked, ignoring my comment.

"Yeah, there were some moments like that," I said, shaking off memories of the *jeune* legionnaire before taking his own life. "There are two types of sins; sins of commission, and sins of omission. I don't know which are worse."

"I regret killing those people but felt that I had no choice, you know?" Taking a drag from his joint, his mouth twitched. His eyes were distant but filled with tears. "My actions ruined many lives, not just my own." He was talking into the air, to himself, or to God but not to me. "My wife, my baby girl, how can they ever forgive me?"

Al Pac's body trembled with the surge of pain that he allowed to surface. He carried a sorrow that I would never understand. His tears fell silently to the ground, and I was the only one to witness that weak moment. We both slept with heavy hearts that night.

After my first week in prison, just when all hope was lost, Anna came one morning to see me with a message from home. Sitting in a narrow visitor's room, I was handcuffed to my chair, and Anna was escorted in by a guard. We were given just five minutes to talk. There were no other inmates in the room. It then occurred to me that none of the prisoners had received visitors since I'd been there. Anna carried an envelope and a small white container. The chair screeched on the tiles as she pulled it out to sit in front of me.

"You're a saint!" I rejoiced, trying to sound cheerful. The last thing I wanted was her worrying about me.

"My own daughter never gave me as many troubles as you! Why didn't you tell them the truth?" she started, her eyes glistening with tears.

"But, Ani, my passport *was* lost…" I flicked my eyes towards the officer in the corner, hoping she would understand my cue. "I will be out of here soon. Don't worry."

"Connor managed to get hold of your sister. She has sent me a copy of a South African identification document. I gave it to the court. Your family is… well, upset," said Anna remaining calm.

"Tell them that I'm okay and that I'm being treated well. It's a shitty situation, but we're sorting it out."

Anna shrugged and handed me the envelope. Inside was a hard copy of an email from my mother and my sister. I read it in front of her, knowing that I wouldn't be allowed to take it back to my cell. It ended with "…Ally, don't try to be a hero. Be honest with your jailors. The truth will set you free…"

I stopped reading and swallowed the lump in my throat. I was ashamed of the destruction that I had left in my wake. I bit my tongue and avoided recounting the horrors I had experienced, while affronted by some fluffy suggestion that I tell the truth? Only a fool could think of love, brotherhood, and truth as he rots away behind bars. I folded the epistle and gave it back to Anna without a word.

"Hey, I brought you some food," she smiled, changing the subject.

Just then, Scholten entered the room. His polished shoes clicked as he walked up to us. Anna instinctively hid her container under the table.

"I have news from your Foreign Legion." The corner of his mouth had curled up into a sneer.

"They have reviewed their records and assured us that you have never been a member; you are a man with no country. They said that your identity card is false and that we can do as we wish with you."

Was the Legion actually keeping its word and trying to protect my identity, even by denying that I ever existed? Or were they leaving a sinner to his just deserts? I wasn't surprised by their response but unsure as to where that left me. I remained silent and let Scholten continue. He pulled up a chair, swung it around, and straddled it back-to-front, sitting between Anna and me.

"We are handing you over to the French since that's where you say your last port of entry was. Tomorrow you will be deported back to Guyane." He stood up, replaced the chair, and walked away. Just before he reached the door, he looked over his shoulder with his signature smirk. "Then you are on your own."

For the first time in jail, I slept peacefully.

"Hey, Al Pac," I said the next morning. "I'm stepping out to get my stuff from town. I'll be back."

I was handcuffed, escorted out of jail to be taken to *The Basecamp*. Again, three armed guards accompanied me. Feeling emboldened now that my release was at hand, I asked them why so much security was needed for a mere illegal alien. The guard sitting in the front passenger seat turned around: "It's for our protection, not yours. You are military trained." I had no idea how being able to march for a week straight made me a qualified killer. They had no idea how their lethal weapon detainee was often petrified with uncertainty and sometimes fear, but I wasn't about to ruin their fantasy.

I hesitatingly directed them to *The Basecamp* and prayed that my mates had heeded my warning the week prior and completely vacated it. Anybody caught loitering would undoubtedly be guilty merely by association. I didn't want to send Al Pac a new roommate. The place was deserted, but the last idiots had left the door unlocked, so we just pushed it open and entered. The dive looked worse than a Dickens-era flophouse. Empty food containers were lying on the threadbare couches, unwashed crockery left on the dining room table, and Parbo Bier cans lay strewn around the overflowing garbage bin in the kitchen. Something was rotting in the fridge. I suspected that the legionnaires who left this mess didn't pass room inspection during *instruction* on the first round.

The guards immediately searched every nook and cranny. "Be my guest," I said, "but you'll probably wish you hadn't." They opened every cupboard, emptying all the contents onto the ground. Th Magazines and DVDs were lying next to the television, which was left on. They shook any seat cushions to see if anything suspicious would fall out and then followed with every mattress. I was lucky to have avoided hanging out with the hard-drug users in the regiment. Serious drunks suddenly felt like a blessing. I was especially fortunate to have refused the Serbian drug lord's offer.

Miraculously, my belongings were still intact and sat at the end of the corridor.

"Do you have any weapons?" one of the guards asked me, as he rummaged through my kitbag.

"Just a jungle knife," I said, knowing they would find it.

"Why? What were you going to do with it?"

"It's part of our military gear," *you idiot*, I said, omitting the afterthought.

My dirty washing had been left in a pile in the bathroom, wet. It was now a living organism, housing a family of frogs. The smell made us all gag.

"I told you," I muttered.

They ordered me to put it all into my rucksack and take it back to jail to wash it, presumably in the turd-water. But I did as ordered, and grabbed my pressed military uniform, and a few books. I wasn't about to get my jungle knife back, as the guard deftly pocketed it for himself. All that I owned now was with me, and we headed back.

As soon as I returned to my cell, I called Al Pac over. He sat down on his bed and me on my hammock.

"Dude, I know you were angry with me the other day. I want to atone for it." I handed him my military jacket and *képi blanc*. "I'm passing this on to you. Possessions are never ours. We simply hold gifts from the ultimate owner." He reached out in silence to take the items from me. His gaze remained fixed on my glowing *képi*. He ran his fingers around the smooth, white circumference, revering it as if it were the Ark of the Covenant. He gripped the flat black rim and placed the *képi* onto his head. When he lifted his gaze, his eyes danced and his cheeks wrinkled into a toothless smile—I prayed he wouldn't try to kiss me. I was moved to see joy fill the worn out face of a broken soul. At that moment the Legion's sacred headdress meant so much more to him than it did to me. He didn't say a word and didn't need to.

Within seconds, he had pulled my military jacket over his head and was marching awkwardly up and down the corridor, shouting orders to the others and saluting me. He was in his element, and I hoped I hadn't just fed a monster.

I left him to his role-play and grabbed a bucket from the bathroom to begin hand washing my pile of algae-infested clothing.

"Legionnaire!" he shouted from the other end of the passage. "What are you doing?"

"I'm leaving tomorrow, Al Pac. I need to get my stuff ready."

He motioned for me to sit down next to him.

"Alex, my friend, let me tell you something. You are my guest, and it would be a great offense if my visitor were to wash his own clothes."

"No, really..."

"Alex," he paused, making eye contact. The *képi* was a few sizes too large, and he had to tilt his head slightly to look at me. It gave him an air of authority. "You don't understand. You are leaving tomorrow. Me, I'm staying here. This is my world. Here, I am *subcommandante*. These men crave discipline and work, especially the kids. If I don't express my power here, I have nothing. If I have nothing, I might as well die. Let me do my job."

In his few sentences, I learned more about Pecking Order Theory than I had during my entire time in the Legion.

Al Pac whistled, and a young Singaporean dropped whatever he was doing and ran towards us. "Wash these," he demanded, holding my rucksack out towards the man's chest, "and find two other men to help you." My personal helper even nodded at me respectfully as he headed towards the washroom. I helped him with my sopping load, and we got to talk. His pawnshop had run into trouble and was on the brink of collapse, so he had decided to do one "mule job" to Europe for a fistful of cash. Unfortunately, one of the cocaine-filled pellets he had swallowed burst before he could board his flight. He had to seek immediate medical attention or die. The prison was filled with the casualties of a flourishing drug trade. For every low-level player who was caught, hundreds who weren't. Miraculously, within two hours, I had dry, clean clothes. I marveled at Al Pac's influence and already missed him before I left.

Early the next day, I was called out of my cell. I slung my rucksack of clean clothes over my shoulder and approached my fellow inmates who were already dressed in their tracksuits to begin the day's training regime—without me. I gave each new friend a warm handshake, a morning tradition I had learned in the Legion. I was proud that I was leaving a semblance of order and meaning in that hellhole, and then I turned my back on them and walked up the stairs to freedom, something few of them would ever experience.

Al Pac still dressed in my military kit, gave me his daily ration of bread and chili as a parting gift. It tasted fabulous!

"Al Pac, this time I'm leaving for real," I said, "Keep yourself together."

"Don't forget, when I visit you, you get me some guns, okay?" he had said as we hugged.

"Ah, yes, good man, if only..." I saw his eyes mist over as he reached over to light his morning joint. "The first step towards freedom is contrition..." I shouted back to Al Pac as I was escorted up the stairs.

In the back of a small police car, I watched the lush rainforest fly past as we drove east towards the border. But I struggled with the thought of leaving the others behind. Unlike the unfortunate Brazilian kid who had committed the same crime as I, I was the privileged white man who was freed. *Why?* As our car negotiated the rocky road towards the frontier, I realized that those answers would be left behind in the Surinamese Amazon. I had a new set of problems facing me.

My escort was to hand me over to the Guyanese authorities. They would most certainly throw me back in the Legion. I expected to spend forty days in prison for going AWOL, where they would break my will to run off. I would learn to love what I previously hadn't, and then life would continue as usual.

I contemplated trying to escape before the handover, but it wasn't wise to take such a risk when I was so close to being released. But going back to the Legion was certainly not an option, which meant somewhere between Suriname and Guyane, I needed to vanish. This game wasn't over yet.

With me in the back of the car was another prisoner, a Guyanese who was also being deported. Unlike me, he looked unkempt and emaciated and was presumably the latest POW of the unwinnable drug war.

Just then, he tapped the side of my leg with his cuffed hands. I flinched.

"How much you got?" he whispered.

I responded with a look of confusion.

"Cash. We can bribe," he said, motioning with his eyes towards the policeman driver. I looked down at my rucksack on my lap. I was also cuffed but remembered that I had three hundred Euros in it. I indicated the number three with my fingers.

Speaking the local Taki Taki dialect, he immediately negotiated with the driver, first offering one hundred Euros for him to stop the car and let us go. The policeman curtly refused. We then doubled-down on the offer, suggesting the officer tell his superiors that the commando and convict overpowered him and escaped. Still, he wouldn't budge. Finally, we went all in—but this time he didn't even look at us in his rearview mirror. It was our bad luck to be assigned the only incorruptible cop in the whole of Suriname.

When we arrived at the border, we were led to a holding cell opposite the customs office where I should have had my passport stamped, to begin with. The post overlooked the river, which was alive with pirogues dashing back and forth like a bask of crocodiles in a relay race.

"You wait here," the incorruptible policeman said to us as he locked us in. I didn't bother asking for an explanation. I was already scheming a plan once on French soil, and I welcomed the extra time to perfect it. The cell was stuffy and damp, and the smell of mold made it feel like a European dungeon. It was dark, except for a thin sliver of light squeezing through the doorframe. My fellow deportee was distressed, pacing back and forth, desperate not to be handed over. Later that evening, as the splinter of sunlight started to burn orange, the door opened, and two officers ordered us to our feet. In tow was a civilian pirogue pilot.

"He will take you across the river to the French authorities," one said, pointing to the driver. I had no idea why on earth they would go through all the trouble of escorting us to the river only to let us be taken across by a layperson. I wasn't about to protest.

We were walking down the embankment to the water's edge before our handcuffs were removed. I gazed across the river, tail between my legs. I had left Guyane with so much exhilaration after escaping the fire-breathing Legion and was arrogant enough to ignore any consequences. We stepped into the bobbing pirogue knowing that I had to make a move before getting to the other side. I didn't want to hurt the innocent pilot to save my skin, however.

Halfway across the river, I addressed the driver in Dutch: "Sir, I have an offer to make you." He gave me, a skeptical look. I returned my best smile. "I will give you thirty Euros if you turn the boat one kilometer downstream and drop me off there." He looked up at the customs office behind us, at the post on the other side, and back at me. Thirty Euros was a week's wages for him. He was torn, and after a few moments, he returned a conspiring grin.

"Okay," he whispered, "but I have a better idea."

In broken English, he continued. "I first park at border post. Me, get out and walk up bank. Police see person getting out the boat. After, I come back and take you down river." He seemed pleased with himself. His master plan to dupe the authorities made absolutely no sense, but I had no choice but to go with it.

The Guyanese deportee jumped at the opportunity. "He is paying for both of us," he declared, pointing his thumb at me.

When we reached the French side, we bobbed up and down in the pirogue while our driver killed the outboard engine, tramped up the riverbank and disappeared at the top of the hill. The sun was setting, and there was a slight chill in the air. The scurry of boats had quieted, as people were packing up for the day. We sat, exposed, vulnerable, waiting. Unlike the movies, there was nothing glamorous or fun about being on the run from the authorities.

Our driver finally returned after an hour, convinced that he'd fooled the officials on both sides. It was a lucrative day for him. He turned on the engine and drove us downriver.

"Only this far. The night is here," he said, docking next to a patch of thick foliage on the riverbank.

"Thank you," I said, handing him the cash.

He left us on the slimy shore, happily waving goodbye. My fellow escapee, knowing exactly how much money I had on me after our earlier failed negotiations, had the balls to ask me for a hundred Euros, to leave me and head his own way. I didn't want any additional company to bring any unwanted attention.

"Here's your bag of silver. Walk right..."

Daylight was fading fast, and I needed to find somewhere to crash for the night. I made my way up the riverbank and towards the road without trouble. The moon had just risen, and an eerie gloom surrounded me as I stumbled through the wet overgrowth into a clearing. I was so elated at escaping that I didn't notice the vehicles until I walked right into them.

A gendarmerie roadblock. *Shit.*

"Stop right there!" an armed member shouted in French. "What are you doing here?"

There were four armed vehicles and six Gendarmes. I wouldn't stand a chance if I made a run for it. No good options lay ahead. They could arrest me and send me straight back to the Surinamese authorities, who would certainly not be this lenient the second time around. Or they could send me directly to the Legion. I decided the latter was the lesser of the two evils.

"I'm a legionnaire, sir. I'm on holiday." I even managed to flash him my best, lighthearted smile, despite anxiety eating my insides.

"Your papers," he said, Gestapo-like, without acknowledging my beam.

My palms started sweating, and my heart moved into my throat. A mosquito landed on my right cheek, hoping for a deep drink, but my hyper-vigilant senses triggered me to slap it before it could draw what little blood I had left. Mustering all my might to keep calm, I reached casually into my rucksack for my ID. In the rush to leave Suriname, I wasn't even sure if my jailors had given it back to me. The gendarme drew his flashlight and inspected it carefully. From his *première classe* chevron, I saw that he was a *jeune*.

"Darklay, I don't know what you're doing, but get yourself unfucked and get back to REI, now!"

TOWER OF BABEL

Proud of your status as a legionnaire, you will display this pride, by your turnout, always impeccable, your behavior, ever worthy though modest, your living-quarters, always tidy.

Code d'honneur

I woke up with a great sense of relief the next morning, well rested in a five Euro per night hotel in Saint Laurent. Like a drug dealer, I began thumbing through my dwindling stack of cash. I stared up at the fan which was hanging by one last screw. Shit, I was still in Guyane.

As I watched the sun streaming through a crack in the hazy windows, my grumbling stomach reminded me that I hadn't eaten in a day. I swallowed my emotions and got out of bed to call Vasile. I gave him my hotel address and promised to stay put and wait for him. Though it was sheer folly for me to step foot again in Suriname, Vasile and I had to rendezvous with Connor there, as he'd be waiting for us with our boat. Pick your poison. Suriname and Guyane were both countries that I had no business being in. I now had little else to do that day other than pull on some jeans and head to the nearby shop to stock up on Kronenbourg.

I had to keep a low profile, though, for the jungle was teeming with my former colleagues. The hours and days passed excruciatingly slowly as I stared at my limited cache of alcohol. I would drink and pass out by lunchtime. Then I would wake up and kill the hangover with more alcohol. My spirit was unwell, tormented by self-doubt, tinged with masochism. I was still in prison but this time, one of my own making. Solitary confinement seemed a more humane punishment than flogging, but I wasn't so sure. If I ever escaped from my room, I vowed never to be alone again. But was I genuinely alone or were there angels around me?

"*What have you done Alex?*" I said out loud. "*Is this your death wish? The Legion is your family. Take your jail sentence like a man, and get back to work.*"

And then I started to reminisce about my lockup in Suriname. I missed Al Pac and the other inmates. I felt one with the universe while I was reclined in my hammock listening to Al Pac talk about God and redemption, which kept my self-doubt at bay. But now I wasn't really alive; I merely existed.

Finally, my de Bruyn backbone straightened, and I focused my energies not on weakening my body and soul with booze but on keeping it fit. In my claustrophobic quarters, I did sit-ups, pushups, air squats, and burpees. I jotted my progress on the back of a pamphlet with a dull pencil. Time seemed to pass more quickly, and my sanity returned—just.

As I turned my room upside down looking for a can opener to gorge on some tinned beans, low and behold, I found not only a Gideon's bible but a box full of books. My curiosity for reading that I had discovered in the infirmary was reawakened. I began devouring every written word, and my mind drifted again to a magical place of prose and poetry. Never before had I appreciated the thoughtful craft of storytelling. I wondered if I would ever pen any of my own tales of adventure and mishap. As a ham-handed rugbyist, I doubted it.

In between Gulliver's travels, I chipped away at some of the more colorful stories in the bible. *How could Jonah have survived in the belly of a whale?* Allegory or not, the moral was that he never lost faith and hope, despite seemingly insurmountable circumstances. In his own existential anguish, English reformer John Hooper, the martyred Bishop of Gloucester, wrote poignantly: "Water is what always has done all the damage."

Yet, as soon as I put down my worn and dog-eared books, my existential funk returned: I was still on the run, waiting. I struggled to distinguish between reality and the written word. But as if an opium haze, I was beginning to enjoy my fantasy neverland. In the midst of Shadrack's torment by fire, he was sent angels to comfort him. Was Vasile one of those cherubs in disguise? After having spoken to nobody but the voices in my head for seven days, Vasile's voice was tranquil music to my ears.

I snapped out of my trance. By the time he phoned, he was already on his way to Saint Laurent. It took me ten minutes to pack my clothes, but it would be several hours before Vasile arrived, so I sat on the grey wingback chair next to the window and soaked in some of the sun, fresh air, and lush foliage.

Finally, I heard a knock on the door.

"One moment. Who is it?" I inquired, trying not to sound suspicious and failing somewhat.

"It's Vasile, your best *camarade*, who else?" came his quiet voice.

I let him in.

"Darklay!" he exclaimed, greeting me. "You look like shit. What'd they do to you?"

"You don't look too hot yourself. Late night with a new girlfriend?" I asked. He smirked.

We walked down the stairs and out onto the streets. I had paid a week upfront for the room so that I could leave without answering any questions.

"We'll talk later," I said to him. "For now, we just need to get back across the border. Anna's expecting us."

"Who is Anna?"

"She's our only friend in the entire country of Suriname. You'll love her curry—and her!"

As incorrigible as Cool Hand Luke, Vasile and I crossed back into Suriname as we had done many times before: without any hassle, and most importantly, without being noticed by any authorities.

Big Mamma Anna greeted me at the door with a slap across the top of my *boule-a-zero* head.

"You stupid boy!" she said, turning to walk back inside. "Have you any idea how much trouble you made? I warned you, cowboys, to be careful, and that you would get caught one day. I told you I wouldn't help you, but now look. How can I not assist you poor fools?"

Tears streamed down her cheeks.

"I had to lie to your grieving mother. And now, you are back, still without a passport. It's too much. I worried about you as if you were my own, stuck in that godless place. Please tell me they didn't hurt you, the authorities or those disgusting inmates."

"Not quite. Sometimes you find angels in the mire and clay."

Still upset, in the kitchen, Anna slammed cupboard doors and threw around utensils to emphasize her point. From the dining room, sulking in our guilt, we boyishly watched her cook.

"I guess that means our curry tonight is going to be really hot," I whispered to Vasile. He sniggered under his breath.

A cheap Bollywood movie played on the TV.

"Boys, sit down now," said Anna, flipping instantly from angry to warm and hospitable. But my anxiety had destroyed my appetite.

"Great. I'm starving, Ani," I lied. One does not refuse an Indian woman's food.

Two hours later, we were all still sitting around Anna's dining room table, buzzed from chili peppers, food, and wine, laughing to stories of Al Pac and the gang. Even Vasile joined the conversation after a few bottles of *Pinard*. Anna's daughter was out for the evening.

For some reason, I couldn't help but notice that Vasile's hand brushed Anna's every time she handed him a beer. As the night came to an end, she arranged a makeshift bed for her charges. Anna had already done so much for us, and I didn't want to take advantage of her hospitality, but we couldn't go out with Scholten running around. Connor was due to arrive soon, and we needed to get the hell outta Dodge.

"Look, dude," I said to Vasile as I unfolded my bed sheet. "You can stay up and do whatever but don't disturb my sleep."

Escaping the Amazon

Connor finally made it back in one piece and phoned us. It was a sweltering afternoon when Vasile and I went to meet him. For some reason, he was desperate for a drink and assured us that we would be safe meeting at a particular local bar.

The dive was smoky, and my eyes needed a moment to adjust to the dim light. Only a few drunks were sitting around. A cranky old radio played Samba, which seemed out of place. The barwoman sat on a high stool, chewing gum and cleaning her fingernails. A fan whirled slowly from a ceiling rung.

I approached Connor's tall figure, slumped over the bar like an old, worn out sailor. When he looked up at me, I barely recognized him. He was ill, his face pale, and his beard had grown thick and matted. His eyes were ten years older and lifeless.

"Legionnaire," I said, putting my hand on his shoulder, "what the hell happened to you?"

He didn't say a word as Vasile, and I sat down tone on each side of him. Vasile ordered a round. The bartender put her nail file down, wiped the sweat off her brow, and pulled three bottles of Parbo Bier from a Styrofoam cooler.

"I am never getting on that fucking sloop again!" Connor started without prompting and filled us in on his travails fetching the craft from Trinidad.

By the time I met Connor in the Legion, he was already at a point in his life where he had little to live for. Few things frightened him, and as such, he seemed to take everything in his stride. On marches, the corporals saw him like a mule, capable of carrying any load. Yet he never once complained, which was why I was astonished to hear him fret, *never again*.

218

We listened as he described his seven days alone at sea. Connor had managed to get to Trinidad without difficulty, where he met the seller who had advertised the forty-four-foot sloop online. A stack of cash exchanged hands. The yacht was a little old but seemed seaworthy. Good enough to get us across the ocean through any storm, the seller assured him.

Connor sailed out of Trinidad quite easily. The sun was shining, and he felt like a proper captain. But the weather turned the following day. Sailing a formidable yacht with a crew of only two is challenging—with no experience, alone, and through a tempest was a death wish. After some time, he noticed that the bow was progressively dipping deeper into each wave, and heeling slightly to starboard. He couldn't understand what was causing it, but something wasn't right. As he was alone, with no self-steering gear, it was perilous to leave the deck unmanned to look below. It was only when his water ran out that he was forced to, and when he did, he saw that the boat was flooded.

The sloop was slowly sinking, but he couldn't figure out where the water was coming from, for there was no noticeable leak to patch. He tried to switch on the bilge pumps, but it was too late since the high water had already killed all the electronics. He ran up on deck and started frantically pumping the manual bilge system.

"It wasn't fucking working. I was going down," he explained as if reliving it.

The storm was still raging, and waves now crashed over the bow, which caused the boat to tilt even more.

"I thought I was done for. I grabbed the life raft and gathered the remaining drinking water as I prepared to abandon ship." I saw Connor's hand trembled, and tension filled the bar.

In a last-ditch effort, still unsure why the vessel was leaking so heavily, he had no choice but to leave the cockpit again to check on the anchor locker at the bow. When he opened it, he saw water gushing in. He then noticed that the winch lever, which had a tennis ball attached to it as a handle, was lodged in the outlet drain hole, blocking the water from draining back out to the sea. He dislodged the ball and ran back to the manual bilge to pump ferociously, as his life literally depended on it.

"My shoulders were on fire, and my fingers bled. My back was cramping. You thought digging trenches in the Legion sucked? Imagine digging ten of them in a row within several hours. After what seemed like a lifetime, I finally got enough water out of the hull to save the ship. I collapsed onto the deck, exhausted and in pain but relieved that I might live for one more day."

Winds on those seas were erratic and unpredictable, and the autopilot was shot. With the damage that the water had caused, Connor had had to helm and trim sails around the clock. He didn't sleep a wink for days. He fell back on the sleepless misery of the jungle course to get him through. But sheer exhaustion eventually got the better of him. He passed out, chipping his tooth on the yacht's wheel, only waking when the boat beached itself on a sandbank, luckily just a few miles from the Surinamese shore.

"*Kurwa*, I could see the lights of Paramaribo but had no idea how I was going to get out of this. So close but so far. I was ready to give up. Why the fuck was I risking my life to save you assholes? All I could do was lie on the deck and count the stars. On the horizon was our Star of the Sea, but I noticed that there were far fewer twinkling beacons in the heavens than usual, and the moon was full. I then totally forgot that I was supposed to abandon you guys."

We had ordered another round as Connor calmed down. We clinked our shot glasses and pushed back the tequila.

"That definitely wasn't a good time to ponder a career in astronomy," Vasile joked. "Maybe you rescuing us was written in the stars?"

"Very poetic, asshole. No, it was the Spring tide that saved the day," Connor retorted. "You two weren't going to come and save *me*, so I put my hopes in a higher power. I passed out again, and it was a miracle that, several hours later, I woke up surrounded by boats and drifting just off the mouth of the Suriname River. It's funny how one only believes in God after He gets you out of a bind."

"True that," Vasile added. "He helped me find my car keys once."

Connor made it, but the vessel was now damaged, as was Connor's morale. Before we knew it, it was well past midnight, and we were the last three in the bar. The bartender was now listening in and topping-up our drinks.

"Let's go see the boat now," Vasile slurred, placing his hand on Connor's shoulder as a gesture of commiseration.

"You can do that tomorrow morning. I'm never getting back on that godforsaken thing again."

"Hey Connor," I interrupted, with an anticlimactic gesture, "did I mention to you that I was locked up? No big deal. Tell you later."

We wandered back to Anna's place and crashed. The next day we were up early, accompanied by a reluctant Connor, to assess the damage on our getaway vehicle and how we could get past the harbor border control. We didn't have much of a plan besides merely escaping into international waters. The final obstacle now was somehow getting Connor on board, by hook or by crook.

We walked past the central market, weaving our way through tourists, local traders, and taxis packed to the brim with luggage. We caught glimpses of the sun rising over the Suriname River as we passed the hovels lining Waterkant Road. I did a poor job of going incognito, wearing Vasile's broad-rimmed "rapper" baseball cap. I may as well have worn a giant sign above my head with my name on it, because, as three tall white men in Paramaribo, it was impossible to blend in.

"Harbor" was hardly the term for the location where Connor had anchored the sloop. The river itself was the harbor; there was no pier or dock, only a haphazard arrangement of boats floating just off the shoreline. It entered Paramaribo from the south, carrying water from the massive Brokopondo reservoir. The waterway then made a sharp turn before curving into a horseshoe to reach the ocean. We stood at the southern rung, looking for Connor's boat. To our right, we could see a long bridge spanning the river towards a leafy suburb.

Very near us, was a large, old, shipwreck resting wearily on a sandbank in the middle of the river. The De Goslar was a German merchant ship purposely sunk by its Nazi crew to avoid it from falling into the hands of the Dutch. It lay there silently, a symbol of man's warlike folly.

When Connor had arrived at the harbor the previous day, he had dropped his anchor and rowed the dingy to shore to evade any questions about licensing or permits. But now we saw that the anchor had dragged, for the boat had drifted from where he had moored it.

Connor was in no mood to help us get to it. I was in no mood to argue or arm-wrestle and thought it best to give him a few days to chill before bringing it up later. In the meantime, I needed cash and remembered that I had some unfinished business with my old pal Jimmy. Like a drug-addicted landlord who'd run out of funds, I didn't just want my rent, come hell or high water, I was determined to get it all.

"I don't have any money," Jimmy blubbered after we forced our way into his casino before it opened for business.

"Figure it out, Jimmy. Time to pay the piper."

"Sorry! I'll figure something out."

"Look asshole! You put me in a situation where I had no option but to give that piece of shit drug lord my money."

"I did you a favor. You should have taken those drugs in exchange. We both could've made a fortune!"

His denial stoked the rage I'd stored up sitting in my cell.

"I just got out of jail down the street. Those bastards searched every square inch of my body and apartment. I'm about to beat the piss out of you."

"Ok, calm down. As you said, you didn't take the drugs, and you're safe now. Listen, I'm working on a deal that will get you your money. Why don't you and your mates come back tonight for a drink so we can talk?"

I reflected on the *Képi Blanc* march in the dead of winter, ages ago and in another world. I carried Marcos' FAMAS when he was hurting and falling behind. It wasn't my fault that I was stronger than the others. In the Legion, those who have been entrusted with much, much more will be asked of them. The last shall be first, and the first shall be last. Weaklings, idiots, and the ignorant skated through life. The Guyanese deportee returned my favor by fleecing me of money when he deserved none. But I walked away free from that sin, still with broad shoulders. Some crosses are simply heavier than others. Jimmy never paid me back, though I did get the crumpled notes he had in his pockets and a potentially authentic Rolex watch.

"Let me tell you how this is going to work, Jimmy," I said that night. "You will repay your debt through work. Every time my mates and I come here, you treat us like kings. We don't pay for drinks. This is our new *Basecamp*, China Beach if you like."

We spent the following sweaty days and nights at Jimmy's casino, drinking, discussing, and throwing out ridiculous plans for our escape. To have any chance of survival, Vasile and I needed Connor. Vasile and I had hardly ever stepped foot on a yacht, much less piloted one. But any skill could be taught, and with or without Connor, de Bruyn was going to make this happen.

"Look," I said to Connor, "you can choose to sail with us or teach me what you know. Either way, that ship's leaving harbor."

Escaping the Amazon

And after the ninth round of drinks one night, I finally convinced Connor to get back on the sloop. Connor had come to think of me as a younger brother—I was his Jannie. His desire to watch over me was greater than his disdain for the vast Blue Amazon. Pushed into a corner to choose between the two, he begrudgingly agreed to come with us. He sulked and cursed for the rest of the evening, but in my heart, I knew he'd pull for us. Connor wasn't just a legionnaire; he was the ideal man encompassing all the beatitudes. There is no greater love than to lay down one's life for one's friends.

"You'll be fine, Connor. Come on. Haven't you read the *Old Man and the Sea*, or *Moby Dick*. The sails go up, and the boat goes forward. We can do that. This is our last thrill to cap off what started when we signed our contract in France. I've always dreamed of sailing from South America to Africa."

My rose-tinted glasses were colored by the fact that I had seafaring in my blood. My parents met sailing around the Cape in the late seventies. Shortly after marrying, they flew to England to buy a boat, which they sailed for six months back to Cape Town before starting a family. If my dad hadn't died a decade later, I'd be a proficient sailor.

When Anna let slip to my mother that we were planning to sail our way home, the professional seafarers in my family, some who'd circumnavigated the world, strongly urged me not to. Winter was the worst season to make such a voyage. Most of all, they questioned the skill-set of our motley trio. But even if the worriers and naysayers were right, I wasn't left with any other options.

"You have no idea what you're getting yourself into," Connor muttered.

A few days later Connor and I were back at the water's edge, having managed to lasso the boat back in so that we could start working on it. Vasile stayed in town to help Anna move some furniture and would meet us after lunch. We suspected that he'd be lifting other heavy objects besides chests of drawers.

Jimmy introduced me to his local handyman, Roy, who worked in the harbor doing odd jobs and fixing boats. He agreed to have a look at ours.

"It just looks like normal wear and tear. We'll have you on your way soon!" he shouted from the bowels of the sloop. But from where we stood, our boat still looked somewhat rough. The mainsail was torn, the sheets were worn and tangled, and there were some new gashes in the hull. Ever the optimist, I took him at his word.

Roy, a diminutive descendant of slaves, a mix of the *Old Man and the Sea* meets Bob Marley. He was a simple middle-aged man who lived in a hut near the river with his wife and two daughters, the quintessential Hemingway character who had known nothing but fixing boats his entire life. He'd scratched out enough to put his daughters—his pride and joy—through school. Roy smiled often, and thick lines formed in his cheeks and around his eyes. He wore his wrinkles proudly, souvenirs of hard years working outside in the elements. He had a firm handshake that commanded respect and trust. I was enamored by his ennobling view of work and the peace he had with his place in the universe. Any Californian or Australian surfer who bragged about his simple beach lifestyle couldn't hold a candle to Roy.

"Crazy young men. Always in a hurry. Now is not a good season for you Europeans to sail across the Atlantic," he said, crouching at the bow, nimbly tying a knot in one of the sheets. I was taking panicked mental notes of which sheets are used for what, and the assortment of sailing knots for each purpose, depending upon location.

"Roy, you only got the second half of our crazy story," I laughed.

"Do not test God or your guardian angels. It's a long way to Africa. You, people, fuss too much about where you need to be. Appreciate what you have now, in *this* moment. You are even welcome to stay here and live with me!"

"I like how you think, Roy, but it's a little more complicated than that for us three."

He gave me his signature broad smile, that of a sage, and lazily lay back on the deck, letting out a big sigh of gratification. "You young men are not content because you are unhappy with life. You keep running, fighting, and searching. Truth and meaning lie right in front of you, but you've been too distracted to notice. Listen, I come out here every day, breathe in the salty air, and stare out at the water. I work to feed my family, and my daughters consider me their hero. At night, I make love to my wife. I sleep nine deep, peaceful hours every evening. There is a small church up the hill that we go to. What more can a man want, Alex?"

"You do your thing, and I do mine," I said, throwing a polishing cloth at him. I suddenly became uncomfortable contemplating the deeper truths of life, especially when I was losing the argument. "Now, are you going to get off your ass and do some work?"

"Of course!" he jumped up, tossing the cloth back at me. "She is a beauty. She's got banged up good. You will need a new staysail, but I should be able to patch up the mainsail for you. Just give me a few weeks. But for now, we take a long lunch. Come!"

He hopped off onto the pier and slapped Connor and me on the shoulders. "Say, guys, I have been meaning to ask you something. My oldest daughter is turning twenty-one next month, and we're celebrating. Please be my guests."

SODOM

Injustice is relatively easy to bear; it is justice that hurts.

H.L. Mencken

Late one night we found ourselves pushing back some free Kronenbourgs at Jimmy's casino terrace, ever watchful that the barwoman wasn't giving us the cheaper local brand. With only four plastic tables sheltered by red and white gingham umbrellas, the place was full. And then Jimmy reared his ugly head. Every time I saw his alcohol-burn face, my anxiety spiked. Nothing good ever came from it. For some reason, he was overly eager to introduce us to Piero, a somewhat odd but gentlemanly VIP guest. Piero was a friendly dandy. He was generous in equal measure. Over the course of the next few days, for no apparent reason, Piero treated us to expensive drinks and lavish meals and gave us free reign to the casino. He even invited us, the whole trio, to his colonial home for dinner. There, we met his lovely wife and two impeccably well-behaved children. Only then did we get to know a bit about this mysterious man's past and what brought such a refined person to this forgotten corner of the world.

Piero was an Italian who had grown up in Zimbabwe, so in a sense, we were cut from similar cloth. "We are both proud European-Africans," he said showing me his rare Krugerrand collection, "and that's why we should look out for each other." Yet despite his outward benevolence, something about my new best friend didn't add up.

The following day the crew continued toiling away on the boat under the unforgiving sun.

"My friends, now it is time for lunch," Roy announced in his laid-back tone. I invite you to my home along the riverbed. Food is cooking now."

As we all sat down at the table and prepared to feast on a sumptuous meal of black beans and fried plantains, I heard a voice that I wished I hadn't. Walking towards us was Jimmy. It was 32º C, and in tow was Piero, in a three-piece suit, a tweed one at that. I greeted them and hesitantly invited them to join us.

"Jimmy drove me here to see how you boys are doing," Piero said taking his seat. "I was simply wondering if you were thinking of sailing again."

"Well, I haven't sailed yet, so to speak. Connor is the experienced one here," I winked at Connor. Roy's wife masterfully balanced six full glasses on a tray and placed them down in front of us without spilling a drop.

Not wanting to reveal too much about the sailing plans for fear of Connor changing his mind, I changed the subject and asked Piero what he exactly did for a living. He took a big, long, slow gulp of coconut water and, placing his glass perfectly and precisely on a coaster, looked up at me.

"Funny you ask," he said, letting out a loud burp. "I was just going to offer you a job."

"Oh really? I don't need a job, but I do need money!" we all laughed, and then I turned to Piero, expecting him to explain himself.

"I won't mince words. You all know that the narcotics trade is booming in South America."

He lit a cigarette with a Zippo lighter. "I help guys like you make a fortune. I'm looking for men who are hungry for adventure and who are not afraid of taking risks. You are the kind of talent I need on my team." He leaned back in his chair and blew out a long stream of smoke, then smiled broadly.

"You see, I used to live in Colombia. Beautiful country! You been there?" We all shook our heads. "Well, in Europe, almost all the cocaine has been diluted. But the high-end customers there—they can tell the difference. I have clients in France who will pay up to one hundred times the street price for a pure, undiluted bag."

Only a man in Piero's position could speak so openly about highly illegal acts.

"Sorry man, we're not drug smugglers," Connor mumbled, through a mouthful, ready to move on to another conversation.

"Guys, just hear him out first," Jimmy interjected defensively. "It's not what you think. It's all completely above board."

"Look," Piero continued, "my farmers, they support their immediate and extended family off of coca leaves. My job is not only to make men like you rich but also to help indigenous peasants feed their children. If I don't buy from them, someone else will, and certainly for far less. I provide them with water, schools, and hope."

"That's very touching, Piero," Connor continued. "Coffee grows perfectly fine in Colombia. And honestly, it'd be hard to give a shit about a farmer's ten kids if I had to spend the rest of my life in a Bogotá prison."

"Ah, but that's the beauty of my plan. The usual drug routes are well known and well patrolled by the authorities. Suriname is a tiny, quiet, forgotten country. Nobody pays any attention to us, much less to a sloop manned by three young legionnaires on holiday," he paused to finish off his drink, leaving his food untouched. "You'll arrive at a specific harbor in Portugal, pass through customs unnoticed, and collect some spending money. You'll rest in a five-star hotel and buy a car to travel east through Spain to visit a certain great-aunt in Nice. She just happens to be connected to my clients who will give you cash for your cargo."

"Sounds intriguing, Piero," said Vasile, apparently considering the offer, "but we're AWOL. Our passports are locked in a Legion vault across the border. Unlike here, we can't just sail to Europe and pay off the officials."

"Ah, yes, I love details! We will talk about that later. I'll arrange for you to live in my lawyer's guest quarters for a few weeks. He can address all of those questions. Come," he said elbowing Jimmy to summon the chauffeur, "I should show you your new home. Oh, and it is near a gym so that you young muscular men can stay in top shape..."

He gracefully placed a large note in Roy's hand as a tip.

The repairs on the boat were burning a hole in our savings, and we were approaching our emergency reserves. Anna and her daughter needed space. As such, I was more than happy to take any offers of free room, board, and spending money. *I would never make a deal with the devil, but I was willing to work with and coexist with him.* As a child, I heard sermons in our church about sharing wealth, with the poor. A few known millionaires humbly bragged about tithing a full ten-percent. A group of stay-at-home wives fulfilled their holy obligations by spending a Saturday morning folding donated clothing for charity. In the situation I was in, I understood how simple it was to be generous when one was rich, how easy it was to avoid sin when one could afford to insulate oneself from its snares.

We begrudgingly accepted. Connor, Vasile, and I still carried with us the legionnaire façade of dangerous men, so we knew that Piero and his friends had a basic level of respect for us, we just weren't sure how far that would stretch.

Roy headed back down to the boat and the rest of us rode with Piero to a small block of flats nearby. The entrance gate opened with a click, and we followed Piero inside to a dark foyer. He led us up the stairs to a newly renovated apartment on the second floor. It was an open plan flat, most likely the lawyer's conference room annex, with a south-facing wall of glass overlooking the river. There were two mustard yellow couches in the middle of the room, dwarfed by the empty space around them, and a white wooden slab that served as both a dining room table and a kitchen counter. Each of us got his own bedroom, which was completely bare but for a single bed, brown curtains, and a small lampshade on the floor.

"Please, make yourselves at home and stay as long as you wish. Let me know if you need anything. *Anything*," Piero said handing us a set of keys before leaving.

Connor went straight for the fridge—it was packed with alcohol, canned fruit, and ready-made meals from Italy. "Awesome, man. I could get used to this!" he said as he tossed Vasile and me a cold Parbo Bier. It'd been ages since we'd eaten any sort of hygienic pre-packaged food. We sat in the lounge suite and enjoyed the view, watching fishermen come in from a day's work. Sure enough, I couldn't avoid talking about the ten-ton elephant in the room.

"Guys, we can't let Piero's generosity persuade us to do something stupid."

"Vasile sounded like he was considering Piero's offer," Connor said.

"Hey," he shrugged, "we have to keep our options open. I mean, we're bleeding money. It was worth hearing his proposal."

"What do you think, Connor?" I asked. "Can we can trust Piero?"

"We can't trust anybody, but it doesn't mean that they can't help us in some way."

"We don't know Piero well, or if he's even the guy calling the shots…"

"Listen, before we get ahead of ourselves, we need to agree on one thing," Connor said, propping his elbows on his knees. "Whatever we decide, it has to be unanimous. All or nothing, mates. Deal?"

"*SELVA!*"

A few weeks later I was sprawled out on the carpet, wishing that the effects of Red Bull would wear off. We'd been partying for two days straight and ended up consuming every energy drink in Jimmy's casino. As a result, I could not sleep for the following two days. While Connor and Vasile slept, I was furiously scrubbing and washing every square inch of the apartment. Come morning, I'd feverishly work on the boat.

"I'm never drinking that shit again," I said to Connor as he woke to find me on the balcony cleaning the windows. "And if I find out that Jimmy spiked my drink, I'm finally going to bust his ass."

As I eventually began coming down from the high, I vaguely remembered a conversation we'd all had the previous night with two other people.

"Do you remember anything we spoke about?" Connor asked.

"Sort of."

"Piero and his lawyer came by. They found a way around your lost passports. We all then considered the possibility of lining the hull of our boat with the cocaine and having a go at it."

I began to doze off but was woken by Connor's sharp tone.

"Hey! So, what do you think?"

I stood up, rubbing my eyes, and went to see if there was any food in the fridge.

"Vasile grew balls of steel and tried to negotiate a bigger cut," Connor continued. "Before he got us shot, I offered to buy the drugs at ten times cost price instead of forty, so that we could pocket a higher margin. But he wouldn't budge and became a little defensive."

"Do you really think Piero's in the mood to bargain? He's not making you an offer. It's an ultimatum," I said, munching on a piece of leftover chicken. "He's desperate to offload some stock before his bosses fit him with cement shoes. Oh, and those bosses aren't the poor farmers in Colombia."

Just then there was a knock at the door. I opened it, and there stood Piero, impeccably dressed in black trousers, with a bright pink shirt tucked deeply behind a gold belt. His two top buttons were undone, revealing dark chest hair, amongst which was nestled a thin gold chain. He walked in with a cheerful smile followed by two bodyguards. "I have good news!" he said with a clap of his hands. "Today you will become citizens of Suriname!"

Connor simply stared up at him. He didn't want to give Piero any sense that we were intimidated, and by remaining seated, he maintained his authority. We were skating on a fragile pane of glass.

"What the fuck are you talking about, Piero?" Connor asked, annoyed at how we were being strong-armed. "And what makes you think you can just barge in here at any time of day?" But he overplayed his hand.

Piero looked at us and opened his arms, inviting us to look around. "I'm sorry," he started sarcastically. "I must have the wrong place. Is this not the apartment *I* am paying for, filled with food that *I have* provided? Wait, this *is* my place. So you better shut the fuck up and let me do whatever I want." His bodyguards bowed up.

I saw Connor's face turn red. Connor rarely lost his cool, but when he did, it was ugly. This was the wrong time for a brawl.

"What do you mean, citizens of Suriname?" I asked, trying to divert the negative energy.

"Meet me downstairs in twenty minutes. I need to go buy cigarettes," Piero said, turning to leave, without looking at any of us.

"That guy's winding me up," Connor let out after the door closed. "Who are those two-bit excuses for bodyguards anyway? Gangsters, my ass. If those assholes want a war, we'll give them one."

"At least we know he's afraid of us," Vasile quipped while shoveling a bowl of muesli down his throat, "but we're still not going anywhere with him."

An hour later we made our way out to work on the boat. But Piero was still waiting for us in a black Mercedes parked in front of the gate. One bodyguard sat behind the wheel, and the other opened the passenger doors for us. We had no choice and got in. As we drove off, the second bodyguard stayed behind.

"Okay, Piero, how about you tell us where we're going?" I said.

He sat in the front seat and hadn't looked at us since we'd gotten in. We kept silent. Connor tried to keep his cool. Vasile remained nonchalant, and I was almost passing out from lack of sleep. After a long stretch, Piero unsuccessfully attempted to lighten the mood. "Guys, relax, you are going to make big money soon!"

We drove in silence for another twenty minutes before pulling up outside a large, immaculate, white palace surrounded by high, green, palisade fencing. It was built in the old colonial Dutch architecture style. The entrance was lined with pillars holding up a triangular roof, bearing an elaborate brass emblem. The Surinamese flag was raised full-mast on the front lawn. Gates opened automatically as we pulled up, and we stopped right at the main door.

A well-dressed young man greeted us. "The Foreign Affairs minister is expecting you," he smiled. "Please follow me." I shot a dubious look in Piero's direction.

"Piero, drive us the fuck out of here now!" I exclaimed. "I just got out of jail and am not even supposed to be in this damned country!"

"My friend, I know all that. Trust me."

I had to trust the ethics code of criminals that no harm would come to me as long as I was of benefit to Piero. We were then led into a large boardroom ornately decorated with abstract art and instructed to sit down at the mahogany table which spanned the length of the room.

After several minutes, the minister himself entered. We stood up, waiting for Piero to introduce us.

"Please, sit, sit. Make yourselves at home. The boy will bring some strong coffee for us." He was warm and friendly, and I felt at ease, despite being in a state of disbelief that the tentacles of the drug underworld in Suriname reached this far. "Let us get straight to the point, shall we?" he continued.

We all nodded, still dumbfounded.

"I believe you had trouble replacing your lost passport, young man?" he said, looking directly at me.

I had no idea what Piero had already told him and was unsure of which response to give. No facts were in dispute, and my life was squarely in the minister's hands. I had to trust that the truth would set me free.

"That is correct, sir."

"And I believe you have the same problem," he said, addressing Vasile, "only you have not run into as much trouble with the police as your friend over here? Oh, and Mr. de Bruyn, I do apologize for Mr. Scholten's behavior. Had I known of your ordeal, I would have personally arranged an early release." His face broke into a big jovial grin that turned into a hearty laugh, which shook his portly body.

"Yes, sir," said Vasile, somewhat less buoyantly.

The help arrived with five cups of coffee on a tray. The minister stirred four sugar cubes into his and glugged some down before continuing.

"Well, I have the honor of giving you some excellent news. You have come at exactly the right time, since I just met with President Bouterse, and he has agreed to offer you both Surinamese passports. You are young, promising, and enterprising men who can be a boon to our emerging economy. We know you do not have many options, and not having a passport means you might face legal difficulties." He paused, confident and calm, to look each of us in the eye.

"What's the cost?" I asked in very broad, coded terms.

"Please don't offend me..." he said, placing his hand over his heart. "This is my personal gift."

"And it's greatly appreciated. We're just wondering if there were any processing fees or special taxes."

"Well, in that sense, you are right, and a smart lad. Of course, nothing in life is totally free. But Piero here will discuss that with you later. I leave all that to the professionals," he said, slanting his head in Piero's direction.

"We need some time to think about it," Connor our unofficial spokesman said, even though the discussion didn't exactly apply to him.

"So," the minister continued, ignoring Connor. "I have arranged for you to collect your passports from the embassy in one week. After that, you come back and we can continue this chat, yes?" he smiled impertinently, in stark contrast to his previously warm posture. He downed what was left of his syrupy coffee and left.

We drove home in silence, none of us wanting to talk within earshot of Piero. He dropped us off at the apartment.

"We talk in two days," he said before the chauffeur sped off.

"I thought I had seen it all, and now we have high-level ministers seeking bribes from simple legionnaires?" said Vasile, clearly amused by what Connor and I understood was a serious predicament.

239

"Anybody in this country can be bought. Anyone."

"Exactly, and if anyone can be bought, then anyone can sell us out," said Connor.

Vasile looked puzzled.

After the red bull induced insomnia had worn off, I slept for 36 hours straight. When I woke the following evening, Connor and Vasile were singing at the top of their lungs. They'd begun drinking that afternoon, and I'd slept right through it.

"Screw this country and all its shit!" Connor shouted, raising his glass to an imaginary crowd. "And screw you, Vasile—I don't even have to stay here!" he said, holding his glass up to him, who in his drunken state wholeheartedly agreed with Connor's sentiment. "And screw the Legion!" Connor continued, "and that fucking boat, and the fucking Amazon, and the…" just then I heard a tremendous crash and shattering glass.

I sprang into the living room. Connor had stumbled backwards and fallen through the glass sliding door next to the couch. He was now sitting in a pool of shattered glass, bleeding and laughing uncontrollably. But the laughter quickly turned into sorrow.

"What the hell do you guys know? You have no clue. What the fuck is the point of existing on this godless planet? It took me this long to realize that the Legion's no substitute for life."

A deafening silence overcame the room. Drinking and getting stupid was *de riguer* for us since we had joined the Legion, even for squared away men like Connor, but I'd never seen him like this. I thought back to Pa and my biological father, my commander, and protector, and how I had never seen them intoxicated, fearful, or out of control. But here, it seemed that Connor was finally cracking up. He carried a story and a past that he had never shared, but every now and then it surfaced.

While the toughest jungle course in the world couldn't break him, weeks of desperation, booze, and the shenanigans of a few cheap gangsters was sucking the life right out of our strongest *camarade*. I realized that the only way to put Humpty Dumpty back together again was to get quickly out of Suriname—together.

Piero showed up the next morning. He had been tipped off by the neighbors that there had been a disturbance in the apartment the night before. He arrived alone but with a visible sidearm. The raw smell of vodka oozed from his pores. Walking straight to Connor, Piero pushed him up against the wall and put the gun in his neck.

"You think you are so fucking smart and can do whatever you want in my home?" The vein on his forehead was pulsing. "I made you a generous offer, and you are trying goodwill. How about you toy soldiers get the hell out of this apartment and bring me my funds by the end of the week. Damn, I'm tempted to say fuck it and just throw you from the balcony now."

We suddenly realized that we were in way over our heads. Deal or no deal, in his mind, we now *owed* him a load of money.

In one swift movement, Connor, who was twice Piero's size, knocked the gun to the floor and lifted Piero by his neck, his feet hanging in mid-air. Our boy Connor was back in full Legion mode.

"You listen to me, Piero. If you ever point a gun at my face again, you'd better pull the trigger, because you won't see another fucking day." Piero struggled desperately and was turning blue. But Connor gripped him tighter and continued. "You are nobody and have no idea who you're fucking with. You come here with your little handgun? My mates at the REI have automatic assault rifles and RPGs. Some nasty former professionals from Corsica and Belfast would love to pay you a visit. And by the way, you have two lovely children, and your wife is really fucking hot. *Est que tu comprende, Kurwa?*" Connor then threw him to the ground, leaving him to splutter and cough his way out of the apartment.

Although Vasile was not one to question authority, he was visibly shaken by what just happened.

"Connor, I've never seen you like this. Are you really capable of killing innocent children?"

Connor returned a blank, almost hostile, stare. His lip twitched before he responded.

"Look Vasile, in the Legion we're taught to obey without question. That's are our higher power. It's not my job to doubt the ethics of a command. If you end up getting your throat slit by these guys and your parents ask me if I did everything in my power to prevent that, then what would I say?"

Vasile turned white as a ghost.

"Was I serious?" Connor then cracked a wide smile. "No, but I had you going!"

It was *reveille* at the Farm again as we hurriedly packed up all our gear and were about to haul ass out of the apartment, *pas gymnastique*. In the taxi, as expected, Jimmy began blowing up my phone. After the tenth attempt, I finally answered.

"I'm done with you, Jimmy," I said before he muttered a word.

"Just hear me out, please. Piero came here and was ready to kill somebody. What the hell happened?"

"Look, all you need to know is that we're out. There's no deal."

"Wait, what do you mean? You can't back out now. You already... I mean, you all lived in the apartment, accepted gifts and met with the minister. That's how deals are made in Suriname, and by that, you agreed."

By the tone of his voice, Jimmy seemed more hurt and betrayed than upset.

"Alex, as you probably guessed, this is for Piero's boss. By you backing down, you are sending a man to his death. Are you willing to live the rest of your life with that?"

"Jimmy, delete my number from your phone and don't ever call me again."

For a moment, I thought back to an ethics course at University. The lecturer began with a classic hypothesis. "From a distance you see a train rolling down the tracks. A person is sitting on these tracks. You have the ability to divert the train to save the person, but in doing so, the train falls off a cliff, killing all the passengers. What do you do?" By the conclusion of the lecture, we were no closer to an answer. I posed the question later that semester to a theology professor. His response was only slightly less confusing. "Consequentialism. Do not commit evil, even to prevent a greater evil. Because at the end of the day, what you'd be left with is still evil."

"Sorry, Ani," we said as she opened the door and we shoved our way into her home. "We shouldn't have left your place."

"Boys, I love you. But you've pushed things too far," Anna said in a nervous motherly tone, quickly whipping up a large serving of chicken butter masala for all.

We agreed that we were in a no-win situation, even if we played along with Piero and the minister. We had to get the sloop ready as soon as possible. Luckily we'd made good progress with the repairs those past weeks, and Roy had assured us that she was at least "seaworthy." But he strongly urged us to replace the fried-out radio, the self-steering gear, and a dozen fuses. He repeated himself several times so as to wash his hands of us should tragedy occur. But we had run out of time for said luxuries and simply had to make do with what we had.

"And now you are leaving me to risk your lives on some broken boat? Don't be so foolish," Anna pleaded.

"Ani, just put all your amazing curry into sealable plastic bags. You won't be far from our hearts and stomachs on this journey."

"Please don't forget me..." she sighed.

EXODUS

Out of sight of land the sailor feels safe. It is the beach that worries him.

Charles G. Davis.

As soon as we reached the port, we remembered that Roy was celebrating the birthday of his 21-year-old daughter. With pissed off mobsters on our tail, this was a bad time to join the festivities, but unlike other times when I had broken my promises, this time I had to be true to my word, damn the consequences. I realized that there were things in life more important than myself. And curiously, that idea was actually liberating.

"Guys," I said. We're going to stop by Roy's first."

They both returned a perplexed look.

"Hey, my boys!" Roy shouted when he saw us from a distance and walked up to me with a big bear hug. He shoved drinks into our hands. "Welcome! We were expecting you all."

"Old man, I wouldn't miss it even if my life depended on it."

I was introduced to his daughter, a beautiful girl with radiant mocha skin. She was decked out in her best dress, probably the only formal wear she owned. But tonight, she was a queen. She flitted her eyelashes at Connor, but out of respect for Roy, there was no way, whatsoever, that any of us was about to go near her.

The other dozen guests were close family and friends, along with a few beach bums from up the coast. The children gathered driftwood and started a fire. And then Roy came in with a net full of fresh seafood. The best Michelin starred restaurant in Paris couldn't compete with the feast we gorged on. We scooped the rice with our hands and picked out the crabmeat with our fingers.

"Roy," I said, in a somber tone. "We're going to be taking the boat out for a bit. Okay?"

I could not specifically tell him that we were leaving for good, and I didn't want to leave him vulnerable to Piero beating the information out of him.

"Yes, my friend, I know that your journey will go well," he responded silently as he walked away.

"Hey, Roy," I shouted before he got too far, "you called us crazy when you met us. You're absolutely right. It's crazy for me to pass up the opportunity to live with you, raise my kids under the shade of a coconut tree, worship in the morning and surf in the afternoons!"

To avoid harbor fees, we'd been docking the boat in a small cove just past Roy's home. The night was cool, and a sliver of the moon gave us a dim offering of light that allowed us to walk along the river's edge to where our sloop rested. All we could hear was the rhythmical slap of water against the hull. Avoiding attention from patrols or locals, we quietly moved our sloop to a nearby dock so we could load it with tinned meat, baked beans, long life milk, cheese, biscuits—and cases of bottled water. Not a word was spoken, for we were lost in our own thoughts.

"Time to start our new odyssey," I whispered to the others.

The sails were still furled, and we fired up the engine. We needed to power ourselves to the river mouth a few kilometers down before we could put the sails up to catch some wind.

"Connor, you take the helm. Vasile, you untie the stern line, and I'll take the one on the bow."

We both gave the boat a strong push, but Vasile accidentally dropped his line. We froze when we heard a loud *thunk,* and the engine died.

"Shit," Vasile said worriedly, still onshore. "The rope got tangled in the propeller." By this time our forty-four-foot boat was already moving downstream with neither power nor sails.

"Run!" I shouted at Vasile, forgetting that we were trying to remain silent. I sprinted along the water's edge and dove onto the boat. Vasile followed but only managed to grab hold of the fender. In textbook jungle course form, I strategically grabbed the back of his trousers, leveraged his sopping wet body, and heaved him on deck.

"The Legion never leaves their own behind," I said gingerly, "even the dumb fucks who can't handle a rope."

"Darklay, we're heading straight for the shipwreck," Connor exclaimed. "I can't get the engine to start!" He furiously began unfurling and raising the genoa, and I set to hoisting the mainsail, in the hopes that we could catch what little wind there was to avoid the inevitable collision. A few puffs came our way, and we barely dodged that bullet, but the sails soon went limp with the strong current pulling us quickly towards the shoreline where there was a small police outpost.

"Vasile, drop the anchor. Now!" I yelled from the stern.

"It's too heavy! Connor, give me a hand!" he roared.

"Hurry guys, we only have one shot. It'd better hold!"

Connor and Vasile lugged the anchor and dropped it over the side of the boat. It took a few seconds before we felt it hit bottom, slowly bringing our sloop to a halt.

247

"Don't get too excited, it might have slowed us down, but we need to make sure it'll hold through the night. We're in a shitty situation and will have to inspect the engine tomorrow morning." I started furling the mainsail again, and Connor followed suit with the flapping genoa. We couldn't be caught inside the vessel if discovered by the police, and so swam to shore and took a taxi back to Anna's place to regroup.

Anna had long ago stopped being surprised by us showing up unexpectedly and did not even bat an eyelid when three water soaked ex-legionnaires arrived at her doorstep at 03h00. We managed to squeeze in a few hours of sleep.

We were back on the boat by the time the golden sun peered over the horizon. I stripped down to my shorts, donned a mask, and dove in to assess the propeller. I clearly saw that the line was coiled around the blades in a tangled mess.

"I need a knife!" I said to Vasile after I surfaced for air. "I think I can cut the rope loose."

Amazonian legionnaires never travel without their machete. Vasile reached into his rucksack and tossed me his. I wedged it between my teeth, took a gulp of air, and vanished into the waters for the longest sixty seconds of my life.

"It's completely off now," I said, pulling myself up over the side of the boat. "Let's pray that the engine starts."

Vasile turned the key, and after a few fits, it came alive!

Connor, who was onshore serving as a lookout, swam back and came onboard. "Something doesn't sound right though," he added, "I'm going to go down to the engine compartment to have a look."

We nervously waited for his response. "Son of a bitch, Darklay!" I heard from below.

I rushed down to see what had happened.

"Turn off the motor!" Connor yelled. "See that small puddle of water on the floor near the engine? Either we have a hole in the hull, which is unlikely, or the engine was damaged, causing a break in the seal."

"I'm not a mechanic and have no idea what you're talking about."

"The jammed propeller might have caused a rod to snap. That means there could be water leaking through the shaft." When we looked closer, we saw that he was dead right. The packing gland had snapped the bolts on both sides, which resulted in a slow leak that could quickly expand and sink the boat.

"We have to get the broken bolts out and replace them."

"Yeah, sure man, whatever you need. We brought a toolbox and drill with flexible bits just in case."

My eyes had to adjust to the darkness again when I headed back below deck. We bloodied, busted, and bruised our knuckles trying in vain to remove the stubborn bolts. Even with the right tools, it would have been a daunting task for a trained mechanic in an air-conditioned garage.

"We need to replace the whole packing gland," Connor finally said in defeat. "The only way we can do that is if we take out the prop shaft."

"We can't do that—the boat will sink."

"Yup. We have to take the sloop completely out of the water. I don't see any other way. We're fucked."

Connor carefully packed the drill back into its box while I stood for a moment, thinking.

"So we're just going to let her sink? That's not how we were trained. We have a wounded legionnaire on our hands, and we need to get him back to life. *Jusquabout a tout prix!* We don't know it yet but have to trust that we were given the tools we need. We'll make this work, Connor."

We left Vasile to watch the ship while we marched quickly to the nearest store of any kind. We found a vendor who sold simple tools and supplies for fishermen. He thought it strange to find two "holidaymakers" desperately banging on his door early that morning. Confused, he allowed us to have a look for ourselves. We bought what was on hand, which wasn't much. For years to come, the shopkeeper would tell the story of two insane *jeune* men who thought they could sail across the ocean in a boat held together with vice grips and cable ties.

Twenty minutes later we were wrestling with the propeller again. Though far less secure, the vice grips served the same role as the bolts. I secured them with as many cable ties as I could fit in that confined space.

"Done! No more leak. We just need to make sure it all holds together," I prayed.

"But if this unfastens at sea, we'll sink," Connor pointed out, matter-of-factly. "And that might be a good thing. It'll save me from taking my own life, because like you, I've become ambivalent about whether I live or die. Shit, with that in mind, let's do it," he shrugged.

We went back up on deck, with an unspoken understanding between us that what we were about to embark on could cost us our lives, yet it bothered none of us. As usual, Vasile wasn't following our discussion but simply following our lead. It pained me to think that we might be leading him, like a lamb, to his slaughter.

"Now we head to sea, maties," Connor said, putting on his best pirate accent.

The engine started beautifully, and we headed towards the Atlantic. We still had a long way to go when the engine died…again.

"Are you fucking kidding me?" I raged as Connor headed down to see what had happened.

"There's water in the engine!" he shouted. "I'm going to bleed it out."

Vasile and I waited on deck, trusting that Connor knew what he was doing.

"Okay! That should do it. *Action!*" he announced again after twenty minutes. The engine started and we chugged along happily again.

"Are you sure about this, Connor?" I eventually asked, voicing what we were all thinking. "It's been nearly three days, and we haven't even managed to leave the damn harbor." I believed in God as little as the others, though occasionally He had helped me find a parking space in downtown Johannesburg. But surely the Almighty had bigger things to worry about than three deserters on a boat. Yet something inexplicable was tugging at me. I wanted to explain myself before the others thought I was chickening out.

"Not that *I'm* saying we shouldn't…"

Just then, the motor cut out again. Once more, Connor went to check. But this time water had mixed with the diesel, and we had to bleed the entire system.

"Maybe you're right, Darklay," Vasile said. "Someone's telling us not to go."

I stubbornly pulled myself together from my momentary lapse in sentimentality and pushed thoughts of divine intervention to the side. "No, wait, our freewill overpowers providence. We're in control here, not the universe." I said proudly. "God didn't simply create us to be His hand puppets."

"Huh?" muttered Vasile.

This was the pride that kills.

"We need to put ourselves in a situation where it is impossible to change our minds," I continued, "like Cortez burning his ships."

Connor came back up from the engine. His hands were covered in diesel, but his face was beaming. "Darklay, you referring to the low tide?"

"Exactly!" I said.

"I don't understand," Vasile replied, confused. "Maybe we are free to decide, but we are still drawn towards good things, not bad things."

"Just hear me out. Approximately three hours before low tide tomorrow morning, the water will rush out of the river into the ocean at five knots per hour. That'll spit us out into the sea in no time. And then we can raise our sails and not screw with the engine!"

"It also means there's no turning back," Connor clarified.

"Better tell Anna to save us a place at the dinner table tonight. Unless Piero and Jimmy have taken our spots."

And so again we had our extra, extra special last curry supper. The next morning, we said our farewells to Anna and her daughter, promising that this would be the absolute last time and that we'd rather die than fail again.

Just before sunrise, the river was draining like a bathtub, as expected. We raised our anchor, kept the sails furled and moved with the tide. Connor was in full Captain mode and was to follow a well-known trans-Atlantic course from Paramaribo to Forteleza, then to St Helena—a speck in the middle of the ocean—Napoleon Island, and finally, Walvis Bay in Namibia. On paper, it seemed as simple as baking a cake. We were so confident that I even arranged for a Namibian mate of mine to pick us up there with cash on hand to pay off passport control.

The morning was magnificent. There were patches of streaky clouds low in the sky, and the sun reflected off them from beneath the horizon. A golden furnace rose above the thick jungle that lined the river. Spanish galleons, French gunboats, and steamers had skirted these shores centuries earlier, pelted by poisoned arrows from hostile tribes when this land was even stranger and less explored. Their crews were fevered, delirious, and longing to return home. *Plus ca change.* I finally felt again the elation from the adventure that I had longed for since leaving the Legion, knowing that we had no radio, self-steering gear, or electricity, an engine held together with vice grips and cable ties, and a crew of one sailor and two glorified cabin boys. We set out to cross just as winter was about to wreak havoc on the Atlantic.

This is it, I thought.

ODYSSEUS

In combat: you will act without relish of your tasks, or hatred; you will respect the vanquished enemy and will never abandon your wounded or your dead, nor will you under any circumstances surrender your arms.

Code d'honneur

In the Legion, I learned how to survive in the jungle by being thrown into the jungle. In the same vein, I learned how to sail simply by sailing. The fear of death was a better motivator than a corporal's fist. Counter-intuitively, I learned that to effectively move in the direction from which the wind was blowing, we had to sail in a zigzag manner to make any headway.

"It's called tacking," Connor had said, ever eager to educate his incompetent first-mates. "Remember marching up the Pyrenees? We also zig-zagged rather than walking straight up." I would far rather have bulldozed my way up those peaks and through the wind, but it just didn't work that way. I remember sitting in Sunday school as a youth, rushing through the lessons, just waiting to get onto the rugby pitch. "I'm in a rush," I said to the instructor. "But God isn't," he replied.

The second nugget that I learned was that even when there was no wind, the craft was never, ever, flat on the water. Like a child's idle toy top, it lists either to port or starboard. This inherent instability became apparent when I suddenly felt all the blood rush from my head. I was overcome with dizziness before bent myself over the side to completely empty my stomach of Ann's curry. I continued dry heaving even after nothing remained. The third thing I learned about sailing was that it was incompatible with my constitution.

Despite my nausea, it was exhilarating to be at sea. We soon crossed into international waters—outside the legal long arm of any country. Vasile and I sat in silence, with the wind roaring in our ears and the salty water splashing up against the side of the boat. I felt free and alive, a feeling that was periodically interspersed with vomiting—every three hours, like clockwork.

Vasile, the least experienced, presumed that he would be most useful if he simply stayed out of our way. For most of the first day, he slept below in the cabin. It was only at sunset when the gusts started to die down, that he popped his head out to join us on deck. It was a magnificent evening; the wind was calm, and the sky was alive with color.

I checked our handheld GPS and saw that we had covered a hundred nautical miles—but only one in the right direction."

"One mile in a whole day of sailing?" Vasile repeated, shocked.

"More like a whole day of *sleeping*, huh, Vasile?" Connor laughed.

"No, one day of nausea," I interjected.

"And now that the wind has died, we might even lose that ground we gained," Connor said. "We're going against the Guiana Current. I warned you blokes before we started," he continued. "The sea is totally unpredictable. I felt a greater sense of certainty on the Farm than in these waters now. And I have no idea how it'll be tomorrow. You take things one day at a time."

As soon as Connor had finished his sermon, Vasile was standing stark naked at the bow, ready for a swim in the calm, blood-warm sea. We had no wind for the rest of the night.

As if on sentry duty at the REI, Connor and I divided the night into three-hour shifts. The wheel had to be manned at all times. But when it was my turn to rest, I couldn't sleep a wink from the incessant vomiting. As soon as I ventured below deck, a dull feeling of weightlessness overcame me. My hands began to sweat, and my stomach lurched up to my throat.

Our second day at sea started under the veil of a thick fog, coloring our world in every shade of grey. The swells had gradually grown throughout the evening, peaking just before daybreak, as had the wind—I knew this because I'd been looking at the water from a painful keeled over position for most of the night.

I mounted the morning shift and helmed the boat, hugging the coastline per an outdated chart we had bought along with the vice grips. The boat's compass was broken. Connor had explained that the best winds were closer to land, but that I should not under any circumstances leave international waters, especially as we approached French territory.

I hoisted the sails again, feeling great satisfaction seeing them fill, impregnated with invisible ether driving them forward. I felt small in comparison to nature's might and yet felt powerful in that I was able to harness it to my benefit. The boat was heeling to port and rocking to the rhythm of the swells beneath her, waves crashing directly against the hull. Lightning crashed far in the distance. The sonic boom smacked my chest before I heard it. The engorged clouds could hold no more water and let out a torrent of rain. We were heading right into the tempest. As if in combat with the Legion, thunder exploded in tremendous blasts around us. It was only a matter of time before a bolt found our high mast an attractive conduit. The noise woke Connor, and he came up from the cabin. He was less perturbed than me.

"Fucking beautiful, man!"

We could barely hear a word through the roar. We crouched forward and tried to shield our eyes from droplets that felt like shards of glass.

"Blessing and a curse! We're hauling ass in the right direction!" Connor lip-synched.

The only way we were going to be able to move ahead was to *use* the storms to our advantage. They were easy to spot, as they hung on the horizon as gigantic black demons, pouring down thick pillars of gray haze. "Let's chase the wind!"

To voluntarily sail directly into a storm was suicidal, but logic had abandoned us when we decided to take on the Atlantic in a broken boat.

"This is how Apollo 11 got to the moon, the slingshot method," Connor shouted.

"Bring it on!" I added spontaneously buoyed by adrenalin as we found our second storm. "This is fun!" A strong gust hit the mainsail on the starboard side, heeling the sloop precariously over on its port hull. But the sails were taut, and the halyards beat fast against the mast. Slowly, the swell increased and turned into breaking waves, attacking us from every angle. We were showered in a waterfall of salt water each time the bow hit one head-on. The boat would then rise and come crashing down like a concrete slab.

We had no life jackets or lifelines, so anything or anybody that washed off the deck was lost forever. My hands, completely exposed to the cold wetness, turned spongy and purple. As if holding a high-voltage cable, my fingers were so frozen to the helm that I had to pry them open just to wipe salt from my eyes.

And then suddenly, out of nowhere, a wall of water towered above us. I looked up, and in one instant knew that if I let go of the wheel, I'd die. The water crashed down on us with all its might. I lost my footing but clung to the wheel, fully submerged in a relentless vortex of violence. I had no time to think, and reflexively held my breath. Eventually, my head found its way to the surface, and my lungs sucked in a glorious breath. I checked on Connor. The force had flung him to the leeward side, and he had smashed his head on the stanchion, to which he miraculously managed to cling on to.

"Connor!" I shouted, trying to get up without letting go of the wheel.

"I'm okay now." There was blood dripping down his face, but he was laughing. "How the fuck did we not capsize?"

This went on for nearly a week. Like Captain Ahab harnessing St. Elmo's fire, our ship simply refused to flip and smash into pieces. We taunted the waves, and every time we survived a broach, our fear of death diminished. We had just stared down the barrel of a FAMAS and lived, and were ready to do it again. "It's only after we've lost everything that we're free to do anything."

"Is that from the bible?" I asked.

"No. *Fight Club*."

"I prefer: *The de Bruyns vow to always do their best*."

"Yeah, I like that better!"

When the winds finally calmed, I suggested that we switch on the engine to keep our momentum. Connor obliged after reminding me that this was not a race. But after a few minutes, we heard a disturbing sound coming from the prop-shaft. I killed the motor, not wanting to risk it shooting a piston through the hull. It seemed futile to investigate and try to remedy the fault, so we simply ceased using the engine altogether.

Connor and I fell back on our Legion training and were able to function without sleep for four days straight. Vasile had slept like a baby through our Man vs. Nature battle. He once sleepily popped his head out in the middle of a storm, and in complete surprise exclaimed: "Hey it's raining!"

Just before I finished my shift on the fifth night and was ready to get some rest, I was mildly delirious. My mind was glazed over with irrational thoughts, and I was barely steering. My brain pounded inside my skull. I was nodding off into micro-sleep. Was I not so exhausted; I would have been able to register that a blinking red light in the distance was not a good sign. But my overloaded mind ignored it.

Suddenly I heard what must have been fish jumping out of the water. I was determined to catch a few to eat. I grabbed my flashlight and shone it on the waves and was surprised to see how muddy they looked considering how far offshore we were. I checked the GPS again, and it showed that we were moving at zero miles per hour—over land. But the sails were taut, and water was rushing past us. There was surely something wrong with the GPS, so I replaced the batteries. But when I verified again, it read the same thing.

"Connor, wake up," I said, shaking him into consciousness. "There's something strange going on."

"Oh shit, dude," he replied as he got his bearings. "We're not moving. You've beached the boat!"

"Impossible," I said. "We're miles from the shore."

Even miles out to sea, we hadn't escaped the Amazon. Unbeknown to us, rainforests regularly deposited enough sand and silt to create vast banks well beyond the charted shoreline. Because I had ignored the red light, I'd sailed directly towards a known hazard. The strong current made it look like we were flying through the water, when in fact, it was flying past us.

"And now we're beached in French waters," Connor added. "When the sun comes up, and the Legion patrols see us, our goose is cooked."

We reefed the sails since the wind was driving us progressively deeper into the sand. We tried rocking the boat to maneuver it free but to no avail. Eventually, we determined that we needed to rotate the sloop enough to at least spin the keel out of the sand. I shook out the reefs and raised the sails, backwinding the genoa on the port side to catch the wind. Connor and I hung our bodies over the windward side of the vessel to control how much it heeled to starboard. There was a huge risk of broaching the sloop, but we had no other options. The wind slowly started driving us over, and the boat tipped towards the water, rotating somewhat. We hung right over the port side to try and prevent a complete broach and prayed for the wind to blow favorably. After two cramped, uncomfortable hours, as the sun rose, we were finally free.

We had no energy to be excited about our success and very soon found ourselves back in the eye of a storm. This time Connor was helming. We had marked the wheel to line up with a mark on the deck so that we would know when the rudder was straight. We desperately needed sleep, and the only way I could stay awake was to try and keep Connor awake. He had suffered a concussion from the tumultuous days before and kept dozing off and hitting his head on the wheel.

"The winds are behaving strangely," Connor slurred. "I'm steering straight towards the back of the storm, but it keeps diverting us away."

I suggested we try and tack as soon as we got near to it, and then come in from a different angle. Connor tried to turn the wheel to port to tack, but it would not budge. I then realized our *jeune* blunder. The wheel was already rotated 360 degrees. While we thought it was lined up in neutral, the rudder was actually at a right angle.

"Fuck this shit! I'm going downstairs," said Connor, "Get that lazy bastard Vasile up here to steer."

"What's been happening, and why's Connor so pissed off?" Vasile asked cheerfully and fully refreshed.

"Just steer."

"Hey, sure. How hard can it be?"

"Dude, I said the same thing a week ago. You have no fucking idea. We need to stay away from the shoreline for a few good days. Then we'll continue southeast towards Africa. For now, just keep us going east."

Like a good legionnaire, Vasile did precisely as he was told and kept the nose of the boat dead East. All I could hear as I tried to sleep was the constant flapping of the genoa, never quite catching much wind, and then flopping to the other side.

Connor and I battled to rest and so checked up on Vasile. Connor's face then turned pale white.

"We haven't moved."

Vasile simply couldn't man the wheel. But Connor and I were now so exhausted that we were as useless as he. And a storm was picking up again.

"Guys, if we keep going like this, we're going to die," I said as we shoveled down some tinned meat. "The plan was to get to Fortaleza in three weeks. At this rate, it's going to take three months."

"Do we turn around?" asked Vasile.

"I'm not making the call. I honestly don't care if I die. Our lives mean nothing in the grand scheme of things," Connor sighed.

Cafard was taking hold of us, and in those situations, murder, mutiny, and cannibalism weren't far behind. Our "safety in numbers" pledge to make decisions unanimously was being tested. Every third thought was of my grave. Pride urged us to announce "every man for himself." Mike Tyson once quipped: "Everybody has a plan. And then they get punched." At that moment, I regained my faith in living and the welfare of my *camarades*. Perhaps deep down, I never really wanted to die.

"God wants us to live—" I put in.

"Just stop!" Connor barked, breaking out of his morbid trance. "Okay, it was incredibly stupid of us to try this in the winter, but we have to stop worshipping death. The Legion taught us to survive, and that's what we're going to do. We're going back to France."

"You mean we're turning ourselves into the Legion back there?"

"No, we're sailing to Europe. We need to get out of these storms and into northern waters. We'll then sail down and hug the coast of West Africa all the way home."

"*SELVA!*"

We'd been slogging through the worst elements for a week, yet the instant we made the definitive decision to go north, the clouds parted, the rain ceased, and the sun came out. Had a higher power tried to prevent us from leaving the harbor in Paramaribo and had the Almighty stopped us from going southeast? It seemed that *a* higher power was now pleased that we had chosen to go north. We were nominal Christians at best, but the timing left us in complete awe. Perhaps due to manly awkwardness, none of us spoke about what was clearly understood by all three. We were witnessing what only Moses and the fictitious Dante saw: the beatific vision. This unapproachable "vision" felt distinct from my previously attainable knowledge of God, *sola fide*. There was beauty in the blue stillness around us. The ocean spread out like a thick blue tablecloth to the very limits of what our eyes could see. The sun floated high above the horizon in a wide-open sky, reflecting off the water like tin foil. Staring at its divine brilliance would kill a mere mortal. In beholding this face, God's creation found perfect happiness.

Now that we were in fine spirits, one at a time, we jumped into the water for a swim, enjoying the thrill of being so far out at sea and not knowing what could be lurking under our bare backsides as we floated face up in the water. We enjoyed a semi-formal lunch on deck, basked in the sun and reflected on the days prior.

"Cheers, to surviving," Connor toasted with a water bottle.

"*SELVA!*"

By the next day, we understood the seriousness of having no tempests—no wind. The clear skies that had initially brought us joy soon turned out to be a curse—we were stuck in the doldrums. The belt of stillness that hugs the earth around its belly was keeping us firmly in place. Our sails were useless if our higher power refused to breathe on us. We were trapped. My first instinct was to accuse God of double-crossing us.

The Blue Amazon held us up under the hot sun as a peace offering—helpless and left entirely to the mercy of her whims. Day after day we watched the benevolent sunrise and creep leisurely in an arc across the sky, torturing us with her slow and controlled metering of time. I remembered past alcohol benders when I would see the sun's rays peek through the blinds and slowly change their angle like the shadow of a sundial. The Legion understood well that nothing good came from a group of bored and confined men.

The powerful sun again took its turn torturing us. We were now running low on drinking water, and all of us had already lost ten kilograms, most of it muscle.

"We're not starving," Connor added cheerfully. "This is the season of Lent, and we're fasting. God wants us to turn those frowns upside down!" Easier said than done.

There was almost no shade on deck, and the hold was a Finnish sauna. But during the day, the boom would cast a slender shadow, and like a faithful dog, I followed it around the mast all afternoon, desperately trying to shield just a fraction of my bare skin.

Connor sported a brutal sunburn but was otherwise back to his old strictly-business self. Neither the sun, boredom, nor insurmountable odds troubled him. Like a medieval monk who hid ancient scrolls from the marauding Vikings, Connor produced a sealed plastic bag full of books. My seasickness had given me blurry vision, and so for hours on end, Connor sat with his feet up, reading out loud to Vasile and me as if we were his doting children. Our Star of the Sea watched over us, as his rich words filled my imagination, and we drifted along with the current.

The next morning as I was assessing how many days of water we had left, I heard Connor shout with joy: "Land! I see land!" He handed us the binoculars and confirmed his discovery with several charts.

"We're going home!" Vasile shouted.

"Welcome to Tobago," Connor said, lifting his nose above a perplexing color map. "It must have been written."

We all looked at each other, high-fived, and performed backflips off the boat. Unable to contain our excitement, we danced, and thumped our chests: "Land!"

We used the last gasps of our rickety engine to pull in to a small landing on Tobago's Southern beach. The water gradually became a pale turquoise, and we could see it gently caressing the milky white sand on the palm tree-lined shore. This was paradise. Hundreds of luxury yachts, immaculate, and state of the art adorned the shoreline. They contrasted with three dirty, burnt, bearded men in a beat up sloop. We had no radio to report our arrival, and nobody cared that we entered Tobago's shores unannounced. We dropped anchor a way out, squashed ourselves into our dinghy, and rowed to the beach.

The moment my feet hit the ground, it felt like the earth was alive and moving. My seasickness returned. "On land? You have got to be kidding me!" Vasile laughed as I struggled to keep my balance. "The only cure for your stomach," Connor added, "is filling it up with lots of beer!"

LAND OF NOD

Every man thinks meanly of himself for not having been a soldier, or not having been at sea.

Samuel Johnson

We felt like aliens who had just exited their spacecraft as we walked through the tree line to a gravel road humming with people and motorbikes. It wound past a row of local vendors selling brightly colored fabrics, beads, mangoes, and freshly caught fish. When we came to a small pizzeria, we jumped at the opportunity. The dive was dark and dingy but opened up into a quaint courtyard where there was a free table in the shade of a coconut tree.

We finished several beers and then wolfed down our pizzas. We continued drinking under the tree until its shade moved away and was replaced by the cool of night. We were well on our way to getting completely smashed. At some point in the evening, between war stories of the massive storms we had survived, the fish we had caught, and swimming with dolphins, we found ourselves entertaining two young women who had come to join our three-man party. Thérèse was a blond French-Canadian; the mocha-skinned brunette, Flor, was Surinamese. They seemed as odd a couple as us three, but they loved our tales, even though we were too drunk to remember much of the conversation.

"So you have secret agent names, were soldiers in that Jamestown Massacre place, somehow escaped by boat trying to get to Europe but washed up on the beach down the road?" prodded Thérèse.

"Yup."

Escaping the Amazon

"Are you sure you all didn't just finish your exams back home and met this morning at the hostel in town?"

Despite our international men of mystery aura and a vague past, they took kindly to us and offered to take us on a driving tour of the island the following day. "Tie me to the mast and plug my ears with wax," said Vasile to me. "I think I'm in love!"

Somehow we made our way back to our dinghy and rowed to the sloop before passing out. It was my first good sleep in ages. Land, beer, civilization, rest—paradise. It'd been less than 24 hours since I'd been battling the seas, and now my mind was filled with sugarplum thoughts of the beautiful tourists nurturing us back to health.

The next morning we rowed back to shore. Thérèse and Flor were patiently waiting for us. In a very endearing gesture, they had brought us egg sandwiches and hot coffee in Styrofoam cups. We munched eagerly and followed them to their hardtop Jeep. We all piled in, ready for the joy ride of a lifetime.

The island was mountainous, and the narrow roads wound in all directions, through the forest and along cliffs that dropped into deep gorges. Inland, we drove through thick, lush, rainforest alive with a symphony of insects, birds, and other creatures. We stopped at all the beaches, each one as stunning as the next. We swam in the sea and played soccer in the sand with the colorful and energetic locals. For once, I wasn't on guard battling the jungle. I could simply appreciate its beauty and God-breathed majesty. Life was good, and I felt my old self again.

Towards the late afternoon, Vasile wandered out on the rocks with Flor, with whom, as expected, he'd already fallen in love. After all our adventures, he had gotten better at talking to girls—I wondered if Anna was actually the one who deserved that credit.

Connor and I sat under a palm tree with Thérèse, and then a few more local friends joined in. Unlike the debauchery I'd seen in the Legion, this was the natural and innocent form of female companionship that I cherished. "Aren't' you glad we didn't die at sea," I said to Connor. "Just look at what we would've missed."

He lay back on the sand and closed his eyes to nap. "Something tells me that it just wasn't our time to die. I still don't quite know how that fate, destiny, and divine intervention stuff works. But I'm open to it."

The sunset infused the sea, sand, sky, and trees with a million shades of crimson and royal purple. But as the evening approached, Thérèse suggested we start heading back to town. So we rounded up the new lovebirds and piled back into the Jeep. The girls sat in the front and the three of us snugly in the back, Vasile on the left behind the passenger seat so he could securely hold his girl's hand as if it were our GPS during a storm.

The fading light flickered through the trees as we rushed past them. We carried on with the talking and laughing as we managed a sharp corner that descended onto a small bridge. We came down onto it at a good clip. In a mere split second, a car coming from the opposite direction took the bend too sharply and slammed right into us. Thérèse hit the brakes, and our Jeep crashed into the guardrail. Our tires barely gripped the road, and we continued skidding down it, metal against metal, with sparks flying everywhere. As soon as we reached the end of the guardrail, we clipped a rung and were thrown into a barrel roll down the steep embankment. All I could see was dust and sky and bushes in quick repeated succession. Although we were belted in, Connor and I somehow came slamming down on top of Vasile. We heard his ribs crack. After what seemed an eternity, the Jeep finally came to a dead stop on its side.

Connor kicked out the side window that was now above him, and we all climbed out as quickly as possible, fearful that the vehicle would burst into flames. We then barely managed to pry the girls out of the mangled wreck. When I was finally able to come to my feet, to my horror, I saw that we were teetering just a meter away from a sheer rocky drop. One more roll and we would have died a horrible death.

The errant driver stood next to his car, still on the bridge, looking down at us in terror. As soon as we were safe, rage built in me like a Caribbean volcano.

"He nearly fucking killed us," I said, and started running towards him, ready to beat him to death. Connor's firm hand caught my bicep, and he pulled me back. I tried to shake him off, but he gripped harder.

"Darklay, we're alive. It was an accident. Drop your stone." But I wasn't listening.

"Let me the fuck go!"

"Darklay, we're here illegally. They'll throw you in jail."

Connor was right, and I slowly got myself together. Other motorists pulled out their phones, and we expected the police to show up shortly.

"We need to get out of here, fast," Vasile exclaimed with a pained grimace, holding his side. "The girls have passports. They're cool."

"Wait. We haven't committed a crime. It was an accident. We can't flee the scene," Connor said, calm and collected.

"Either way," I put in, "the girls are shaken up badly, and we can't just leave them. Come what may, I'll take that risk."

We nervously waited for the authorities, not knowing if we'd be arrested. But when they arrived, we were immediately put at ease. They looked like they had just come from a family crayfish boil on the beach. Clad in short-sleeved, open collar shirts, shorts, and leather sandals, they were surprisingly easygoing—peace, love, and Bob's music. They took Thérèse's statement but didn't ask any other questions, and even gave us a lift back to our boat.

"Whatever life has thrown at us," I said, "somebody really doesn't want us to die. I think that means we truly need to get home."

And as fate would have it, I ended up as close to any South African Embassy as the ocean currents could have taken me. With one in nearby Trinidad, it was foolhardy for us to risk sailing to Europe and then Africa. We were out of harm's way, and there was only a fine line between foolishness and martyrdom. Mark Twain quipped that the latter covered a multitude of sins. I accepted that I was only meant to float to Tobago. Now, all we had to do was get to Trinidad as soon as possible.

We said goodbye to Thérèse and Flor, pulled up our anchor, and used the coastal winds to sail across to the neighboring island 83 kilometers away. There were enough gusts to get us around the western side of Trinidad, and we headed for Port of Spain on its western coastline. Able to finally take our time to enjoy the sailing experience, we navigated our way past several rocky cliffs protruding out of the water. As soon as we approached the harbor, though, the wind died. We were too far out to row in, so we tried the engine one last time. But doing so was her final *coup de grace*. She wouldn't be missed. A toothless fisherman on the shore saw us being pushed towards the rocks and came to our assistance.

Disembarking was bittersweet. As our first day in Aubagne, it was again every man for himself, as we scrambled to get to our respective offices. We had left the unifying brotherhood of the Legion behind and were once again individual foreign nationals. Connor went to declare himself at the customs office on the docks. "Just me," he confirmed to the officer. Vasile vanished, and I wasted no time getting to my South African embassy only a few blocks from the port. The relaxed culture of the islanders was evident even amongst the embassy personnel. Like the policemen in Tobago and the helpful fisherman, the officials greeted me like a long-lost friend. They even remembered me from my first phone call after being arrested and issued me a letter to certify my existence. What a feeling to be on South African soil again and under the protection of the Rainbow Nation!

It would take several days before either Vasile or me had passports. The following evening, Thérèse and Flor showed up and made those last days even sweeter. Vasile disappeared with his soul mate, leaving Connor and me to clean and sand the boat. Connor refused to allow us to sell the old sloop, even though the extra cash would have helped immensely in getting us home.

"I'll be back to fix her one day," Connor said nostalgically.

As it was Connor's swimming winnings that had bought the vessel in the first place, we didn't have a leg to stand on and left him with his dream.

We scraped our last cents together to buy a one-way ticket out of there. And yet we didn't want to go. We had everything we wanted in life on that island—except for permission to stay. Old Roy would have been proud of our conversion. But just like the Legion, Trinidad wasn't a substitute for life.

Finally, the day of our departure came. Vasile was on the verge of tears saying goodbye to Flor. They had on last dance to imaginary music only they could hear and looked like synchronized ballroom dancers. "Look," he said, "everything now is perfectly ordered."

Connor was booked on a flight to London, his first stop, before heading home. Vasile was flying to Romania via Barbados and Frankfurt. I'd accompany him on the initial two legs, as it was the only way of getting home without needing visas.

My worst fears were confirmed when I inspected my new passport for the first time. It looked like a child's construction paper and paste project. It literally consisted of a few folded papers stapled together with my photocopied signature cut and glued from my initial application.

Vasile asked me to take his rucksack off the carousel while he went through Trinidadian customs. I waited while he was held up at the desk. Ten minutes went by, and then twenty. Finally, I saw the officer walk Vasile into a special room. Shortly after that, the same official came out and approached me.

"Passport, now! Whose bag is that?" he asked, pointing at my feet. "You are not allowed to touch another person's bag in the airport."

"But I'm traveling with him. It's his."

"Oh? You are traveling together?" His eyes narrowed, and he put his hand on my shoulder. "You need to come with me."

He led me past the customs desks and into the same room as Vasile. He sat silent, and I had no idea what they'd already discussed. The plane that was to take me to safety was only meters away, and I expected that my freedom was about to be denied at the last minute. Another officer entered the room and threw my passport on the table. "It's fake!" he shouted. "Both of their documents are fake. They are smugglers." He summoned enforcement on his radio. I'd had enough of biting my tongue and turning the other cheek. I wasn't about to endure another injustice.

"Look, I have no weapons. I have no drugs. I have nothing but this outbound ticket and this passport." I waved my documents in his face. "If I don't get on that plane, I'm going to be stuck in your fucking country, with no fucking money, and I'm going to be your fucking problem."

The customs officer phoned our plane to make sure we were listed as passengers. He then called Frankfurt to confirm that we were permitted to land there. After another twenty minutes, he finally gave up. "Gentlemen, I don't know why you came to Trinidad. We don't even know how you arrived here, but please, just leave."

We were randomly searched, checked, detained, and questioned all the way home. We were now probably on every no-fly list and Interpol database—testament to the fact that even when one leaves the Legion, one never, ever actually leaves the Legion. It's a good thing that we had never agreed to smuggle Piero's stash—we made for incompetent criminals.

With our plane in clear view from the bar, we sat for one last drink. It was hard to imagine that I was finally going home. Sometimes freedom felt like slavery. After experiencing a succession of adventures that never entirely took me to where I thought they would, this journey had run its course. Unlike the idealistic *engagé volontaires* of yesteryear, we were a more somber and reflective lot.

"There's no work in Bucharest…" Vasile sulked.

I thought of Connor, who would have been halfway home by then. *One day I might actually know who you are. I won't press you. But for now, let's rest. See you soon, brother.*

My mind remained silent.

A verse that I remembered from Sunday school came to my lips: "*A time to be born, and a time to die; a time to kill, and a time to heal; a time to weep, and a time to laugh.*" I cracked a smile.

THE PRODIGAL SON

Seek in reading and thou shalt find in meditation; knock in prayer and it shall be opened in contemplation."

Saint John of the Cross

Benno and I thought it best to celebrate my survival in the only way we knew how by downing drinks at the local club.

I lied to myself that I was doing so in memory of Jannie, but to me, it was simply a good excuse to get drunk and rowdy. We'd already pushed back a number of shots when a party girl I barely knew asked me to help her. She knew I was a legionnaire and presumed I was the perfect guy for the task of telling another patron to stop harassing her. In textbook macho White Knight mode, I was more than willing to save a fragile damsel in distress and hopefully get a free fistfight out of it.

I made my way through the crowd and approached what looked like a gorilla disguised as a man. He had jet-black hair, gelled into spikes, and Amazonian pythons for arms. Legionnaires were a scrawny lot, and I'd seen corporals with toothpick arms flatten recruits twice their size. Muscles didn't intimidate me.

"Hey mate, see that girl over there? I'd like for you to stop touching her."

He stepped back a fraction to look at me, clenching his jaw.

"Mate, I bumped into her, and she threw her drink at me, threatening to have my ass kicked by some military guy. Keep her in check. She's the sick bitch looking for a fight."

"I don't like your attitude."

"Look, if you really wanna fight, let's fucking take this outside." he challenged me.

Never backing down from a duel, I downed my drink and followed him out. His scrawny friend accompanied us and pleaded sincerely with me to cool down and walk away. When I turned my head to look at him, momentarily off guard; my opponent slammed a hard fist to the side of my face. It dropped me down onto one knee, but I wasn't out. With my face stinging, I looked up and saw him running away up the stairs back into the club. *Fucking coward*, I thought.

The scrawny friend that remained behind was panic-stricken. He apologized and tried to defuse the situation. Exercising his only option, he threw a half-ass punch, which I instinctively dodged. In that instant, months of pent up rage boiled over, and I started violently beating him, merely because he was there. Benno pushed his way past the onlookers to pull me aside and stop me from killing the bloke.

With my shirt torn, nose bleeding, and my lip starting to swell, I sat down on the pavement and tried to get myself together. With the adrenaline fading, I suddenly felt horrible, scared, remorseful for what I'd done and ashamed of the monster lurking inside me. The rest of my mates came out to pick me up.

As we were waiting for the taxi, from out of nowhere, my muscle-bound opponent approached me once more.

"I didn't mean to beat your friend," I said offering him my open hand. "Let's call it a night."

Muscleman took my wrist, gripped it tightly and suddenly pulled me forward, slamming his head into my nose. This dropped me again, and the last thing I saw was a retro, size 12 Air Jordan flying through the air towards my face. An expensive Gucci loafer would have been more dignified. After watching many sacks of potatoes fall in the Legion, it was now my turn to play the sack. This wasn't a Hollywood movie where the unhinged war hero returns home and beats up an entire gang of Hell's Angels.

But the bacchanalia continued. The dark laser accentuated clubs became my man cave. The electronic music lulled me into a blissful meditative trance. They reminded me of the dancehalls in Guyane and Suriname. But now, instead of calmly enjoying a couple of drinks, I found myself drinking ten, and forgetting much of the night.

Weeks later, Benno and I ended up boozing hard at a friend's party back in Stellenbosch where I'd last seen Jannie. Benno spent the evening between the couch and the toilet. Alcohol consumption accentuated the sadness of not having Jannie with us. Legionnaires drink to forget, and then they forget what started them drinking in the first place.

A neighbor showed up at the door and politely asked the host to turn down the music. I watched the two argue for several minutes. When it became physically heated, my Legion fight or flight instinct kicked in again.

"You need to get the fuck out of here," I said to the neighbor getting between the two. The cauldron inside me started to boil over, but this time, I couldn't stop it.

"What the hell are you going to do about it?" he responded.

At that moment, nuclear fission was in full swing. I kicked him in the center of his chest with all my strength. He flew several meters down the passage. I then picked him up, Legion style, and threw him outside like the morning garbage.

"That," I spat, marching back inside and slamming the door.

I grabbed another beer, and the party continued as if nothing had happened. But soon, somebody was pounding on the door again. I opened it and came face to face with the same neighbor.

He was smirking at me. "Ah, you wanna hit me again? Do it!" He launched forward with a right hook, but I ducked. After missing, he was momentarily vulnerable. As my right fist traveled through the air, it carried with it the full weight of Jannie's meaningless death; *Caporal* Bordon; the violation of Myshkin; the desperation of the illegal miners trying to feed their families; the face of the *jeune* legionnaire who ended his life; Al Pac's hopeless future; every hollow shell of a man I had seen in the Paramaribo prison, and the shame of desertion. What my fist didn't carry with it was a semblance of freedom, healing or forgiveness.

My knuckles connected with his jaw, sending his front teeth slicing through his upper lip. He grabbed his face and fell to the ground, blood streaming through his fingers. At that moment I realized that: *If I didn't sort myself out, I would either kill somebody or get killed.*

Two years after the casual conversation with Jannie and Benno that led me to join the Legion, I was the same man, yet I wasn't the same person. Vredendal—*Valley of Peace*—was an appropriate name for the place I was traveling to in hopes of "refocusing." There was no sit-down family intervention, and I wasn't bundled into a van. I went on my own accord to take a break from life after taking a break from life. "*Of course, you're my boy.*"

My family owned property there, on the banks of the Olifants River, in a quiet, deserted area of the Western Cape. There stood an unoccupied farm with no water or electricity, only a barn for shelter. One of the few items I brought with me was a perfectly intact but unread bible from my youth. The great Saint Augustine, a young troublemaker like me, grew up to say:

> *Men go abroad to wonder at the heights of mountains, at the huge waves of the sea, at the long courses of the rivers, at the vast compass of the ocean, at the circular motions of the stars, and they pass by themselves without wondering.*

As such, I found myself unintentionally engaging in the ancient practice of monasticism, seclusion, and withdrawal from the world. Many who served in the Legion also saw it as a monastery of sorts but for the damned. For several weeks I slept in the barn, grew a beard, and ate very little. Every morning, I'd walk several kilometers to a waterfall to fetch water and take a cold shower, enjoying the quiet isolation. Even after the Reformation, the idea of growing in faith through contemplation still held sway, from the formation of Christian communities to organizing retreats—the interior life.

As I walked back to the barn one morning after gathering firewood, I reminisced about how my entire life was consumed with chasing the wind that was always just out of reach. I felt that I had let the world down. I remembered my student days struggling through a course in psychology. I was failing and sought help from the lecturer, one of the few who cared enough to offer it.

"Make a detailed outline of the class based upon the reading and lectures," he suggested.

"Can't I just copy the notes from the others?" I protested.

"No," said the man of letters. "The learning is in the making. You won't be tested on material that wasn't presented to you. Everything you need has always been right there in front of you."

Weighed down by my thoughts, I sat down on a rock, overlooking the empty land and read a random verse from my bible. I hoped to chip away and finish it in a year or two. There was crispness to the air as the winter chill had begun. The grass was dry and dull, and most of the flowers were dead, but there was beauty in the barrenness. The tranquility was consuming and unavoidable, yet I had always disliked stillness. Almost imperceptibly, the wind picked up. It swirled around me, bringing life to the dying vegetation. It was whispering something. I couldn't quite make it out at first, but slowly, hesitantly, my heart listened: *"After the earthquake came a fire, but the Lord was not in the fire. And after the fire came a gentle whisper."*

The breeze helped me understand how even in the most godforsaken places on Earth, people like Al Pac can dream and be human, how the dolphins' light-show gave me the will to survive, how Roy's kindness and generosity could change even the hardest of hearts, like mine. In silence, with brilliant clarity, I now knew that I was *still* living my adventure. Nothing had ended, and no door was closed in the continuum. Every moment, exciting or mundane, was being woven together into the divine mosaic of existence. For the first time in my life, I no longer needed to chase the wind or the storm. Just as we had let go at sea, I could let go now. Ultimately, the Master who commands the winds, jungles, and the waves was the author of my unexpected story.

I found myself sitting in front of the owner of a boutique Finance Firm in Johannesburg. I sat upright in my pressed navy suit. My fingernails were trimmed, my hands flat on the table before me, and not a whisker was unshaved. He glanced over my résumé, confirming dates and degrees.

"Thanks for coming in," he said as he placed his phone on mute. "Let's see. You have a gap of time in your résumé. How about you tell me what you've been doing for the past two years?"

My mind immediately filled, not with sorrow but with joyous memories. Jannie appeared, standing alongside the legionnaire who took his life, both smiling and glowing with light. I relived the pride of finishing top of my class. I recalled the thrill of cheating perdition in the Blue Amazon, and the love I had for Connor and Vasile, who laughed and cried alongside me. Through the majestic sunrises, I revered the living saints who suffered with me, and I loved the divine hand that gently steered us clear from death. Most of all, I saw the God's beatific vision, *paradis*o

"Well, sir, where do I start…"

AFTERWORD

Connor returned to Australia and began a lucrative career as a derivatives trader. He settled into married life and fathered a boy who will one day sit on Connor's lap and hear his version of this same story. We remain in close contact, as I owe my life to him. True to his word, Connor returned to Trinidad to repair our banged up boat. We both promised to revisit there with our families to sail much of our same escape route in the sloop that allowed us to live another day to take pleasure in our beautiful families. He remains an avid swimmer.

I lost contact with Vasile over the years. He struggled to find his calling in Romania for some time, though he is now believed to be serving in the London police force, still protecting the lives of a foreign nation's citizens. One can take a man out of the Legion, but one can never take the Legion out of a man.

Anna still lives in Suriname, alone. Her daughter subsequently moved to Holland but frequently returns to visit. We remain in contact, and I long for her amazing curry dishes to this day.

Daniels stuck out his five-year contract in the Legion and served with distinction in several theaters, whereafter he returned to the American South and married his lifelong love. Always the simple, low-tech man, focusing on family and work, he largely abstained from social media and modern methods of communicating. As such, staying in touch remains a challenge. I look forward to reuniting with him over a drink and hearing his tales of earning the title "Five Year Johnny."

It's springtime in South Africa as I sit comfortably in my Johannesburg home behind my laptop. Feeling the breeze through my open window, I am at peace with my experience. After settling back into "Civvy Street," I soon realized that working for a firm wasn't entirely to my taste and so returned to my entrepreneurial roots. I married a woman who accepted me as whole, with scars, bruises, and all. We were soon blessed with a baby boy. I pasted these small snippets and paragraphs into a story for whoever might care to listen. The burning desire to suck the marrow out of life still boils in my heart—it drives my continuing adventure of everyday life. *I can do all through Him who strengthens me.*

Made in the USA
Middletown, DE
01 December 2024